TOOLS OF THE TRADE

Firefighting Hand Tools and Their Use

Richard A. Fritz

FIRE ENGINEERING®
PennWell Publishing Company

Disclaimer

The recommendations, advice, descriptions, and methods in this book are presented solely for educational purposes. The author and publisher assume no liability whatsoever for any loss or damage that results from the use of any of the material in this book. Use of the material in this book is solely at the risk of the user.

The opinions expressed herein are those of the author, and do not necessarily reflect those of Fire Engineering or PennWell Publishing Company.

Published by Fire Engineering Books & Videos
A Division of PennWell Publishing Company
Park 80 West, Plaza 2
Saddle Brook, NJ 07663
United States of America

EDITED BY JOHN J. NAPOLITANO
BOOK DESIGN BY MAX DESIGN
PHOTOS BY RICHARD A. FRITZ
COVER DESIGN AND ILLUSTRATION BY STEVE HETZEL

2 3 4 5 6 7 8 9 10 11

Printed in the United States of America

Library of Congress Cataloging-in-Publication Data

Fritz, Richard A., 1957-
Tools of the trade : firefighting hand tools and their use /
Richard A. Fritz.
p. cm.
ISBN 0-912212-62-4
1. Fire extinction--Equipment and supplies. I. Title.
TH9360.F75 1997 97-3661
628.9'25--dc21 CIP

■ Rick Fritz has always wanted to be a firefighter. In 1973, he was one of the original members of the Fire Explorer Post organized with the Penfield (NY) Volunteer Fire Company. In 1976, he enlisted in the Illinois National Guard as a crash-rescue firefighter. A year later, he became a career firefighter with the Muscatine (IA) Fire Department and subsequently moved to the Davenport (IA) Fire Department in 1981. In 1985, he completed his Fire Science Management degree through Southern Illinois University. In 1990, he left the fire service to pursue a teaching career at Scott Community College in Bettendorf, Iowa, where he was the department coordinator for the Hazardous Materials Technology program. In 1993, he returned to the fire service as a full-time staff member of the University of Illinois Fire Service Institute in Champaign, Illinois. Rick is very active in fire service training. He develops and delivers firefighter training throughout the State of Illinois. He serves on both IFSTA and NFPA committees, as well as various statewide training committees.

This book is dedicated to my parents and to my wife, Cathy.

ACKNOWLEDGMENTS

I would like to thank the following fire service professionals for their assistance. Without their help, this project would have been impossible: Bradley G. Bone, lieutenant, R.N., Champaign (IL) Fire Department; Michael N. Ciampo, firefighter, FDNY Ladder Co. 44; David F. Clark, firefighter, M.Ed., Illinois Fire Service Institute; Robert E. Farrell, Fire Hooks Unlimited Inc.; Andrew A. Fredericks, firefighter, FDNY Engine Co. 48; David Fulmer, Illinois Fire Service Institute and captain, Savoy (IL) Fire Department; John P. Grasso, firefighter, FDNY Engine Co. 48; Gary Gula, firefighter, Champaign (IL) Fire Department; Craig A. Haigh, chief of department, King, NC; Robert Hianik, Lt., Itasca (IL) Fire Department; Ray Hoff, battalion chief, Chicago (IL) Fire Department; Robert Hoff, battalion chief, Chicago (IL) Fire Department; Rick Kolomay, lieutenant, Schaumburg (IL) Fire Department, and Ladder Company 26, Ladder Company 36, Ladder Company 56 of the FDNY; William R. McGinn, lieutenant, FDNY Engine Co. 48; Robert McKee, battalion chief, Chicago (IL) Fire Department; Andrew O'Donnell, director of training, Chicago Fire Department; Ray Palczynski, lieutenant, Davenport (IA) Fire Department, Savoy (IL) Fire Department; Mark Trujillo, firefighter, Denver (CO) Fire Department & T-N-T Tools, University of Illinois Fire Department; Marty Vitale, president, Iowa-American Firefighting Tools; Victor Vitale, battalion chief (ret.), FDNY, consultant, Iowa American Firefighting Tools; Barry Wagner, facilities manager, Illinois Fire Service Institute; Matt Weber, battalion chief, Urbana (IL) Fire Department; Jeff Welch, lieutenant, Urbana (IL) Fire Department; Mark N. Wesseldine, firefighter, FDNY Ladder Co. 58. Also, a thanks to Dwight Hart, captain, Davenport (IA) Fire Department, wherever you are.

A very special thanks to Craig A. Haigh, Ray Palczynski, and Dave Fulmer for all the reading and assistance!

TABLE OF CONTENTS

"Some guys are so busy learning the tricks of the trade, they forget to take the time to learn the trade."

—Anonymous

A sign with those words hangs in a friend's office. I read that sign and realized that I was one of those guys. I was always looking for a shortcut; something to make the job quicker and easier, although not always better. I thought about what that sign said, and it changed my attitude about a lot of things that I had been doing as a firefighter and a training instructor.

Firefighting is dirty, backbreaking work. We have been combating fires since the beginning of time. While the job has changed a lot in several thousand years, in many ways it has stayed the same.

Much like the military, the fire service has made great technological advances in the past 50 years. Better machines, computers, materials—all sorts of things make our fighting forces more powerful. Like the military, it all still boils down to one fundamental element: the fighter on the line. The military depends on the infantry soldier; your department depends on you.

To be an effective firefighter, you must know how and when to use the tools of your trade. These tools are as important to a firefighter as weaponry is to a soldier. Without them, the enemy is going to kill you.

Our tools may be similar to the tools of other trades, but they are used differently and must meet different standards. Consider this: Painters use ladders, but would you use a painter's ladder? I thought I knew how to use the tools of my trade. In researching and writing this book, I found out how little I really did know. I talked to firefighters all over the country about tools and how they should be used.

Hand tools are critical to all companies on the fireground.

As you will notice in your reading, the most efficient tools available to the fire service have been designed by firefighters—those who know the trade and what we need to accomplish.

Neither reading this book nor watching a video nor buying all the best tools will make you a better firefighter. You always have to practice what you learn. You must know how to use your tools, how to maintain them, and what their capabilities are in the event that you have to push them past their limits. Getting dirty is the only way to learn this trade.

Practicing with hand tools is the only way to become proficient in their use.

Allowing apparatus manufacturers or salespeople to specify and equip your apparatus is insane. Tools and tool specifications should be as critical a factor in designing and buying apparatus as the pump, tires, aerial, or any other part. Don't underestimate yourself or your abilities. Who knows better than you what tools you need?

Firefighting is all about size-up. Size-up must include what tasks are to be performed on the fireground and the best methods of getting those tasks accomplished. Size up your own department first. It's a waste of effort for firefighters to work with tools designed specifically for plaster and lath in response areas where there is only gypsum board. Know your response areas and what tools you need to have. We do this all the time now, but we seem to quit when it gets to hand tools. Hose layouts, crosslays, fittings, adapters—all are planned and specified to the *nth* degree. When we get to hand tools, we are happy just to transfer them from the old rig or the storeroom to the new rig without really giving a second thought to their capabilities or importance.

WHO SHOULD USE THIS BOOK

Firefighters. I wrote this book for several reasons. First and foremost, I wrote it to pass on information I have learned from firefighters to other firefighters. Traditionally, firefighting skills have always been passed along from older, more experienced members to the next generation beginning their careers. Somewhere along the line, that has stopped happening. Veteran firefighters are leaving their jobs and taking their skills and knowledge with them. Younger firefighters are unable to learn from them. The number of fires is down, and there isn't a text, film, or video that can teach the next generation these unique skills.

This book was written to help you learn new skills or hone the ones you already have. It is not an absolute authority; the techniques and methods that I describe are not the only ways to use these tools. They are, however, the best methods I could find based on the experiences of many firefighters, not just my own experiences or opinions.

Company officers. Yesterday you were on the operating end of the tool; today you're telling others what needs to be accomplished. Many departments do not adequately prepare company officers for the critical role they play within their organization. I wrote this book to help you pass on the skills and knowledge you have, combined with the skills and knowledge of some of the best firefighters I know.

Chief officers. During an exhausting training session one afternoon, a comment was made by a chief officer who was watching the drill. The comment struck a nerve with all of the training instructors who heard him make it. As we watched firefighters making an attack on the fire, including throwing ladders, stretching lines, and making hydrants, the chief officer turned to a group of us standing outside the building.

"Ya know," he said, "I'm glad I came down and watched this. Sometimes I forget how long it takes to throw a 35-foot ladder."

Sometimes we all forget. Firefighters are working their hardest to accomplish the most that they possibly can, as fast and efficiently as they can. As chief officers, you need to stay current with the job. It's tough because there are so many other commitments, but this one needs to be at the top of your list. You must continue to support the line firefighters, providing them with every possible advantage.

Trustees, city managers, mayors. Providing fire protection to your constituents is a costly affair—you need to get the most for your dollar. Firefighters can't stand by and let buildings burn or people suffer. They give you everything they have; they've given you their lives. Buying them the most expensive modern tools isn't always what they need; buying them tools that work is!

CHAPTER 1:
CUTTING TOOLS

Manual cutting. It's a very basic skill—a task repeated over and over in recruit school and seldom practiced again once we get on the line. Think about your own community. How many times have you been asked to make a cut? To cut someone out of a car, to cut a chain or padlock, to cut through a wall or fence, to cut into a roof, to cut debris out of the way? Cutting and cutting tools are and always have been closely associated with the fire service and firefighters. Our reliance on power tools has caused our manual cutting skills to deteriorate to a deplorable state. In many cases, our cutting skills are bad because we weren't really taught correctly in the first place, or it was just assumed that we knew how to use an axe.

In this chapter, we will look at some basic cutting tools and cutting techniques. Nothing fancy—no roaring motors, no screaming hydraulic pumps, and no hissing compressors. The discussion will be limited to hand tools driven by your muscle. Hand tools are highly efficient when maintained and used properly. Also, hand tools always start.

Power tools have their place on the fireground, but they should always be backed up by hand tools. Unlike power tools, hand tools always start when they are needed.

SIX-POUND PICKHEAD AXE

"I don't know!" cried the distressed citizen. "They came from all over the place! Funny-looking hats, sooty faces, and they all had axes!"

Heard it before? It's the standard firefighter joke—a fireman with an axe. There is no other tool so closely associated with our trade as the six-pound pickhead axe. It has been a part of the fire service for centuries, and next to the pike pole or battering ram, it's probably the oldest tool we have.

Six-pound pickhead axe.

The six-pound pickhead axe, ancient and venerable, is the most inefficient tool that we currently maintain in our inventory. Six pounds is just not enough weight to get the maximum efficiency from your swing with any kind of measurable result. Firefighters tire more readily with the six-pounder than with a heavier tool. Take those six-pounders off the rig and give them away at retirement dinners, or clean them up and use them for musters and parades. Just in case that isn't possible, let's look at the standard use of the tool.

Standard Use

The pickhead axe was originally designed for use on wooden sailing ships. Sailors used it for fighting fires as well as enemies. The blade would cut through decking, exposing hidden fires, and the pick could be used to start a hole. Lines could be severed, and tangled rigging could be hooked and dragged away. It was also used as a weapon, commonly referred to as a boarding axe. With the advent of the steel-hulled ship, navies worldwide gave away these tools by the thousands to local fire companies. It made sense to clear the storerooms of a tool that was intentionally designed for one thing: fighting fires.

The pickhead axe is primarily a cutting tool, and a cutting tool only. It is used for opening up a burning structure for entry, rescue, and ventilation. Its effectiveness depends on proper maintenance and how good the operator is at swinging it.

To cut properly with a pickhead axe takes practice—lots of practice. A fire axe is different from the wood-cutting axe out in the shed. You must hold it differently, swing it differently, and be safety-minded in the extreme. The accuracy of your swing is important. The more times you can hit exactly where your previous swing landed, the less time you will spend completing the entire operation.

Practice holding the axe and getting a proper stance. Your stance will depend on the type of footing that you are able to get. You'll hold the axe with a different stance if you're on the ground, on a ladder or a roof, in the bucket of a platform, or swinging away inside a structure. Most often, try to get a stance with your feet apart and your body weight centered. Hinge your knees just a little to give you some flexibility. Hold the axe handle where it feels comfortable, but not up close to the head. Your stance should give you balance, and it should feel natural. You must be capable of rocking back and forth on your feet while swinging, at all times maintaining your balance against gravity, the force of the wind, bad weather conditions, and most importantly, any changing fire or structural conditions.

To swing your axe, always remember that you are not Paul Bunyan trying to deforest Oregon. Axe swings should be no higher than your shoulders. Swing the axe in an over-the-head motion, but maintain control of it. On the backswing, the head of the axe should come to a point just slightly behind your head. In most cases, you should be able to see it in your peripheral vision. As you swing downward, slide your upper hand along the handle to meet your other hand, which is firmly grasping the bottom of the handle. Don't try to push the axe or you will tire quickly—allow the weight of the tool to do the work. Let the axe head drop into place at a slight angle to the work being cut. Concentrate on accuracy; it is very important that you angle the blade to the surface, because a dead-on strike may bounce back at you and drive the pick between your eyes.

If you encounter a hard surface, flip the axe over and use the pick to get a good purchase. Just as with the blade side, do not swing the pick over your shoulders. Let it fall accurately and penetrate the surface on its own. If the pick should get stuck, carefully push or pull the handle to the left or right. The pick is designed to free itself if you twist it 15 degrees or more in either direction. The same holds true for the blade. If it gets stuck, don't pull on it or you'll fall. Flex the handle sideways, then rock it up and down to free the blade.

Practice swinging the axe both left- and right-handed. Fireground conditions may not allow you to swing it the way you normally would. Be prepared to swing on your weaker side. Set up an old telephone pole behind the firehouse. Practice batting lefty.

Special Uses

The six-pound pickhead axe's special uses are really pretty limited. It can perform several functions other than cutting, but as we will discuss in later chapters, there are other hand tools that are better suited for multipurpose use.

Its legitimate uses include:

- To force inward- and outward-swinging doors past their latches.
- To force double-hung windows open by prying them in their centers and pulling the lock screws out of the wood.
- To break glass.
- To make a purchase point in automobiles for hydraulic tools by prying or poking holes in the sheet metal.
- For searching.
- To remove moldings, baseboards, and other trim during overhaul.

This six-pound axe is too light for the job. The firefighter must swing harder and more often to accomplish his task.

In-House Modifications

There isn't too much you can do for the six-pound pickhead axes. They are too light to be really effective. If you must use them, see the section under Eight-Pound Pickhead Axe to learn about modifications you can make to the handles to help them be more effective and efficient.

Limitations

Throughout this section, we have discussed the limitations of the six-pound pickhead axe. To review, those limitations are:

- It's too light.
- It cannot be used as a striking tool.
- It is more tiring than a heavier version.
- It's an inefficient cutting tool.

EIGHT-POUND PICKHEAD AXE

Okay, so what's the big deal? A whole new section on the same axe, except it's two pounds heavier? Yes! Two pounds of weight added to the head of a pickhead axe makes all the difference! From reports I have received from firefighters across the country who use the eight-pound pickhead axe, *this* is an *axe!* If you don't think that two pounds can make a difference, take the firehouse challenge. Next time you're at the station, sit down in the kitchen and eat six pounds of ice cream. If you are still conscious after six, eat two more pounds! I think you see my point!

Standard Use

The standard use of the eight-pound pickhead axe is very similar to that of the six-pounder, with one major exception: It's two pounds heavier and therefore requires some slight changes in your stance. It also offers more flexibility in cutting.

Unlike the six-pounder, the eight-pound pickhead can be used very effectively while you are sitting on your butt. A unique technique for opening a pitched roof safely is a technique called the zipper cut. Most

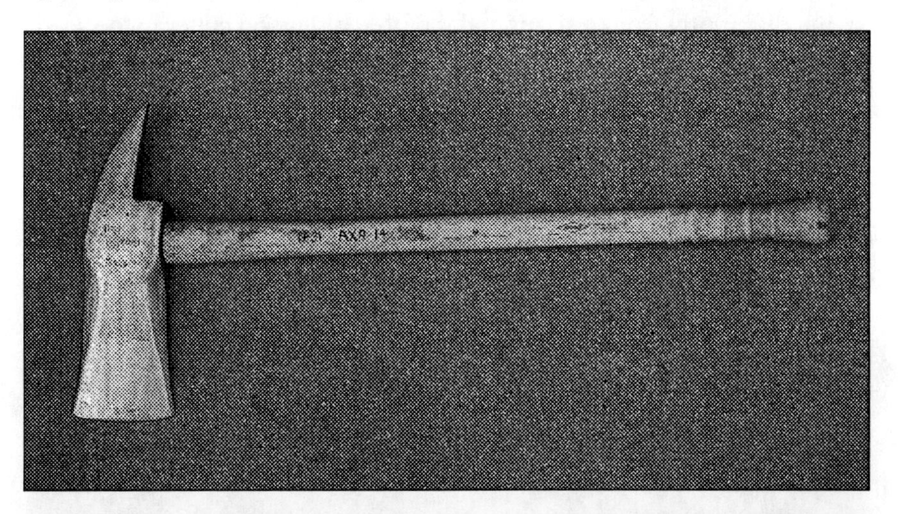

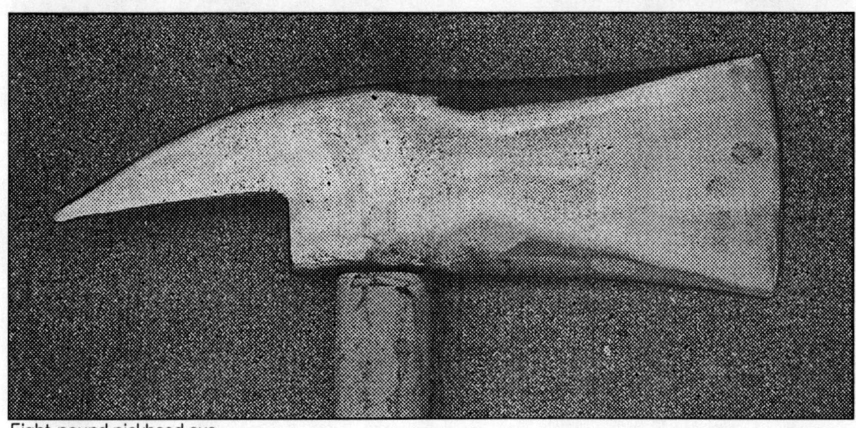

Eight-pound pickhead axe.

firefighters do not enjoy working on pitched roofs, especially in bad weather.

Get an eight-pound pickhead and get on the roof. Move to the ridge beam. Once you have selected the proper location for the hole, straddle the ridge. Sit with one leg on the side you are going to cut, extended straight out at an angle to the roof ridge. Tuck your other leg as if you were riding a horse. To make the cut, swing the axe along the line made by your extended leg. Cut away from your leg. Once you've made about a two-foot-long cut (that is, two feet from the ridge line down the roof), slide backward along the ridge and make a cut parallel to the ridge, two feet down from it as you go. Look where you're going when you move backward. When the second cut is about eight feet long, extending from your original, make another cut like the first along your leg. Slide back along the ridge a little farther and use a pike pole or trash hook (minimum length of 12 feet) to pull the debris out of the hole. Reverse the pole and push down the ceiling. You've never had to stand up, further reducing the chance of a fall.

The stance and swing when using an eight-pound pickhead axe are almost identical to that of the six-pounder. Just be aware that the added two pounds of weight have increased the size of the axe head, making it swing differently from a six-pounder or that old wood cutter in the shed. Be prepared to shift your weight around to accommodate the heavier

Roof operations should be performed by experienced personnel.

head. Pleasantly, you will find, this axe does the work. As before, accuracy is the key. Avoid those Paul Bunyan strokes. Shoulder high, allow your upper hand to slide down the handle to meet your other hand, which has a firm grip on the bottom of the handle. Allow the weight of the tool to do the work. When it gets stuck, free it the same as you would a six-pounder. Push or pull the handle to the right or left to enlarge the cut a little, then rock it free. Pulling will cause you to lose

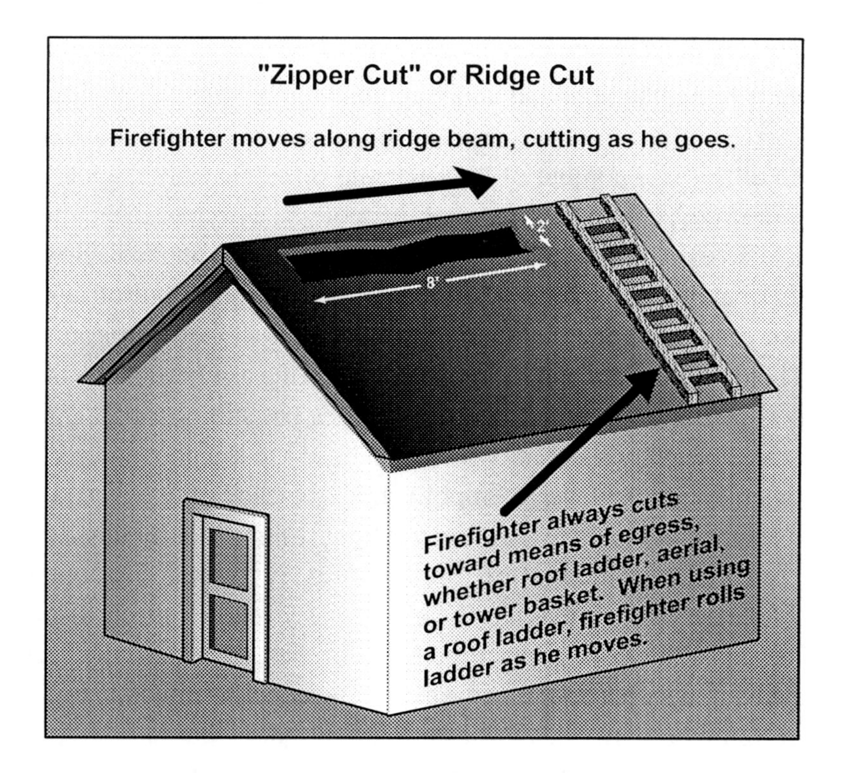

your balance and fall. Practice. Swing both right- and left-handed. Practice flipping the axe around and driving the pick into the surface to be cut. Always swing so that the head strikes the surface at a slight angle. Dead-on strikes are dangerous and a waste of your valuable energy. If the axe bounces, it could spoil your good looks.

A two-firefighter axe team is a very effective and efficient method of creating vent holes. Armed with two eight-pound axes, the firefighters mark out the area to be cut and then begin their work at opposite corners of the marked-off area, working toward each other. When working in this manner, team members can complete the hole faster and can continually see each other, allowing for safer operations.

The eight-pound pickhead axe is a very effective and efficient tool when used correctly. In a strictly technical sense, it is a limited-use tool, designed primarily to be a cutting tool. It does have other limited uses, however, that make it somewhat versatile if it happens to be the only tool in your possession when certain tasks must be performed.

Special Uses

The eight-pound pickhead axe can be used effectively in certain forcible entry situations and overhaul operations. It can also be used to some extent at automobile fires and accidents.

In forcible entry situations, the axe can be used as a wedge to try to force a wooden, outward-swinging door in a wood frame to slide past its catch and pop open. By inserting the blade of the axe into the small space between the door and jamb and then using the pick as a handle, you can apply pressure to spread the door away from the jamb. Hopefully, it will pop out of the latch and open. Be careful about any pressure you exert on the handle of the tool while prying. Even a well-maintained tool may snap if inappropriate pressure is applied sideways to the handle. The axe is designed to take straight-on, not lateral, pressure. It is really embarrassing to come out to the rig for another tool holding a broken handle in one hand and the axe head in the other. This is harder to do with fiberglass handles but still possible. During most forcible entry applications, the pickhead axe is a poor choice. It is not a striking tool. Never attempt to strike the back of a pickhead axe with another tool to drive it in. Get a cutting/striking tool for this.

A quick access method of forcible entry is to try to pry open a double-hung window that has the standard swivel lock mechanism in the center of the sashes. Insert the blade of the axe beneath the center of the bottom sash. (You may have to remove the storm windows first. If so, remove them all. Do not allow any glass shards or framing material to remain.) Pry the handle of the axe upward. The window will begin to open, pulling the wood screws out of the lock, allowing the window to open normally. Be very careful: If the window is not in good condition, the sash will distort and the pane may break, showering you with glass.

The axe can also be used to breach wooden walls, of course, but the heavier eight-pounder can also smash its way through cinder block very effectively. Start the breach by using the pick end to disintegrate some of the block. You may find that the pick is all you need. Swing the axe like a golf club as much as possible to conserve your energy. If you need to use the blade side for a wider striking surface, use it. You'll have to recondition the tool afterward, but that is not an important issue during an emergency. *Note: Using the axe in this manner is misusing the tool.* Do so if it is an emergency—it will save your life! If it isn't an emergency, you are better off selecting a more appropriate tool.

During overhaul, the axe can be good for opening walls, floors, and so on. The blade is a very effective tool for sliding behind baseboards, door trim, and moldings. The pickhead will not allow you to pry off the baseboards easily because its curvature is too great. To use the pickhead to remove baseboard, let the axe slide down across the surface of the wall. Nine times out of 10, the axe blade will wedge itself behind the baseboard. If you let it slide hard enough, it may knock the baseboard right off altogether. The amount of pressure required to remove it will depend on how well it was installed, but generally you will be able to get it off with the axe. Remember, you are applying unnatural force to the axe handle when prying. If whatever you are trying to remove resists, move the tool to a new purchase point. Overapplying pressure may snap the handle of a wooden tool, and if you're into something really tough, you could even break a fiberglass handle.

The pickhead axe is very efficient at removing standard doors from their frames. In most residential and light commercial structures, doors are only attached to their frames by hinges screwed into a wooden stud. To remove a door with little or no damage to either the door or frame, insert the blade of the axe between them, either just above or just below the top hinge. Shut the door. Exert slight pressure on the handle and pry the door away from the frame. The screws will pull out. Open the door, flip the axe around, and repeat the process at the bottom hinge. The door can easily be removed. If there is a center hinge, remove it last, since the door will be less likely to get stuck that way. This is an excellent method for getting doors out of the way during firefighting operations and overhaul. Once removed, a door can be used to cover holes in floors. Little damage is inflicted on either the door or frame.

If you have well-maintained tools, the handle of your eight-pounder is also effective for removing lots of plaster and lath in a short time. While overhauling, open up a decent-size hole three to five feet off the floor. Insert the handle of the axe into the bay. Grasp the blade and pick and pull toward you. The handle will pull large amounts of plaster and lath from the wall. Do this only with a well-maintained tool. If not—*snap,* no handle.

Even though the eight-pound pickhead axe is really a limited-use tool when compared with other hand tools, it does have a variety of uses. The key consideration in using and sometimes misusing the tool is what task you must accomplish. Is it an emergency, or would you be better off waiting for a more appropriate tool? What condition is the tool in? Size-up is important, not just in overall strategies on the fireground but also for simple tasks such as using tools.

In-House Modifications

We'll consider both the six-pound and eight-pound pickhead axes in this section, since these modifications can be done to both tools to make them more efficient. Keep in mind that the eight-pound pickhead is preferred. It is difficult for even the handiest firefighter to come up with modifications for the pickhead axe. It has basically remained unchanged for more than 300 years, but there are some little things you can do to improve it.

The first and foremost in-house modification you can make is good maintenance. Refer to Chapter 9, Tool Maintenance, for suggestions.

At this point, we will discuss nonmaintenance changes to the tool.

The axe is designed to be a cutting tool. To be effective, it must be able to cut through material without binding. Paint on the head will cause it to bind, so remove all the paint from the head of the axe. Either sandpaper or a wire wheel will do this. Polish the surface smooth and shiny. If you really want to go overboard, you can polish the head to a high sheen, but that isn't really necessary. Get a good, smooth surface on the axe, and keep it well oiled with either light motor oil or machine oil.

Protect the handle of the axe with overstrike protection. Some fiberglass handles come with a rubber bumper that slides up the handle to act as a collar just below the head. These are great, but there is an alternative method that works as well and can be more easily replaced if it gets worn or damaged.

Using either 12- or 14-gauge wire, wrap the handle from just below the axe head down about four or five inches. Make the wraps tight. Once the wire is in place, wrap it with black electrician's tape; otherwise, use colored plastic tape if you want to color code the tool. The tape will help hold the wire in place, and it looks good. You can place a second wire wrapping over the top of the first wrap, coming down from the head about half the distance of the first wrap for double overstrike protection. This simple, inexpensive modification will provide more than adequate protection for the handle and help to prevent those embarrassing and potentially dangerous axe-breaking incidents. The overstrike protection is easily changed when worn out and doesn't interfere with the performance of the tool.

Tape fingerholds can be added to the handle to help you get a better grasp when using the tool to pry off trim and other moldings during overhaul. Using an axe as a prying tool is technically a misuse of it. Prying can be done, but you must be careful to work within the limits of the tool and the handle. These finger grips will also help you define the position of your hand in relation to your swing. The tape will help you gauge how high you are swinging.

To measure where to put the tape, hold the tool in a position as though you were going to pry off some molding about waist-high. Look at your hand closest to the axe head. Where your hand wraps around the handle is where you want to put four wraps of friction tape. This will make separators for your fingers, helping to prevent your wet gloved hands from sliding down or off the handle. Make sure the distance between the tape wraps is large enough for a gloved hand. Wrap the tape several times, making it fairly thick. Use friction tape rather than plastic tape, since the idea here is grip, not looks. In general, the fingerholds will probably be about one-third of the way down the handle from the head. These may need to be changed often.

Another modification that can be made is to improve the grip on the axe handle. The idea is to create a grip that will give you something to hang on to with wet firefighting gloves. One way is to wrap the handle with French hitching. This is a spiral-type grip that will give an improved hold. It has been used in navies all over the world for centuries to improve grip on ships wheels, handrails, stanchions, and the like. There are several ways of doing this.

The first method is the most time-consuming. It involves a few yards of cord (thin clothesline, twine, or small sash cord), friction tape, and several hours. First, measure how high up you want the grip to be. It should be at least 18 inches from the bottom of the handle up toward the head. Wrap the handle of the axe in friction tape. This will hold the cord in place. Next, secure the end of the cord to the bottom of the handle with a wrap of plastic tape. Make it tight so it won't pull out. From there, begin making a series of tight half hitches around the handle, one right after the other. Continue this, ensuring that all of the half hitches are made in the same direction. After about three inches or so, you will see the spiral beginning to appear. Keep tying those half hitches! Tie them the entire 18 or more inches until the handle is completely wrapped and the spiral hitching is apparent. It will take you several hours and a few yards of cord to accomplish this. Once done, secure the end of the cord with plastic tape. Now, wrap the entire area you just corded with friction tape, pulling tightly and pressing the tape in place over the cord and spiral. This will actually create the sticky surface of the grip. The big drawback to this method is the time it takes to wrap (and unwrap any mistakes made during the process). It is a major task to replace this type of grip when it wears out. It looks great, so you might want to add it to the grips of those six-pounders you're having chromed. Forget that last wrapping of friction tape—just leave the knots showing. This looks great for parades.

An easier and more effective method of producing the same result is to use electrical cord, sash cord, small-diameter rope, or 14-gauge wire, plus friction tape. Measure out the size of the grip you want as before. Wrap the handle with friction tape. Secure one end of the material to the handle with plastic tape. Make sure it is tight. Carefully wind whatever material you are using (cord, wire) around the handle in an even spiral, like a barber pole. Press it onto the friction tape as you go along. The friction tape will help hold it in place. Secure the other end to the handle with plastic tape when the wrap is finished. Wrap the entire spiral with friction tape, pressing it into place. Tuck it in, over, and around the spiral. It's done. You may want to wrap it twice with the friction tape. Try holding the axe with gloves on. If it doesn't seem to be enough, use a thicker material for the spiral material, or wrap it a couple more times with friction tape.

Now you have an axe that has been modified to help you use it more efficiently. Don't forget to read the chapter on tool maintenance. It's one thing for the tool to look nice, but it must be sound!

Limitations

The eight-pound pickhead axe is truly a limited-use tool. In your inventory of tools there should be others that can do what the pickhead axe can do, plus more. It is a good tool, but recognize its limitations, which are:

- It is heavy.
- Its primary function is to cut.
- It cannot be used as a striking tool.

Because of the weight and size of its head, its value as a forcible entry wedge-type tool may be limited.

BOLT CUTTERS

Standard Uses

Bolt cutters are another tool available to the firefighter for cutting purposes. Designed primarily for industrial use, these provide a quick and relatively easy method of cutting through various materials.

Much like the pickhead axes, there are a variety of other tools available to the firefighter for cutting materials such as chain, lock shackles, fencing, and other such items. Bolt cutters, however, are relatively cheap to purchase. When used properly, they are fast and efficient.

The most important factor in using bolt cutters on the fireground is to have the right set for the material you want to cut. It may be necessary to have different types on the apparatus to cut different-strength materials. The cutting surface may not be suitable for case-hardened material like certain chains and lock shackles but will be good at cutting fencing, light locks, or small cable.

Do a survey of your response area. Is it high-security? Are there lots of chain-link fences? How and what are people using to secure their residences and commercial buildings?

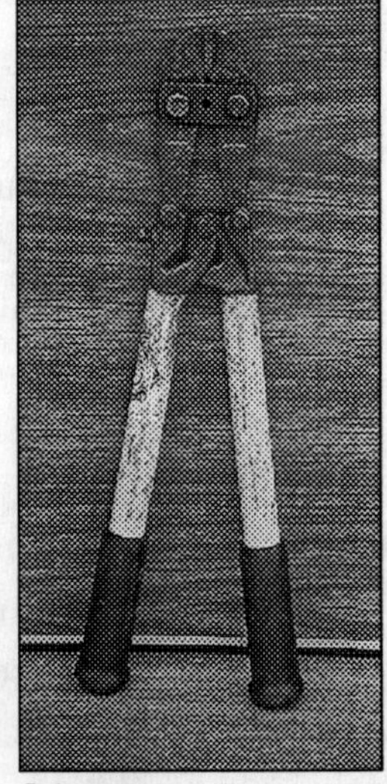

Bolt cutters.

Answering these questions will help you select the right type of bolt cutter, or bolt cutters, to carry on the apparatus. The size of the cutter, as well as its cutting capacity, is very important. Long-handled cutters provide better leverage for the firefighter doing the cutting. Buying a set of cutters just because it fits into a storage compartment doesn't make good sense.

Don't waste money buying dielectric bolt cutters or wire cutters. Firefighters don't cut power lines, period. The bolt cutters should be of high-quality carbon steel, with long, powerful, preferably fiberglass handles with rubber grips. They should have handles long enough that the cutting process is made easy. The shorter the handles, the less leverage you have for cutting.

When using the bolt cutter, size up the material to be cut. Ensure that you have a proper set of bolt cutters capable of biting into and cutting completely through it. *Make sure you wear eye protection!* Cutting with bolt cutters may launch the cut-off end of chain, bolt head, or whatever else you're working on. Protect yourself from flying debris. Never intentionally cut something that is a loose end. When cutting cable or some other material, make sure of the end result before you cut. Cutting cables or cords may release an object being held up or in tension, such as a garage door spring.

In forcible entry, when cutting lock shackles, cut high up on the shackle. Cutting too close to the lock itself may jam it if you don't cut all the way through or if the cutters twist in the process. Cutting high on the shackle gives you another place to get a purchase if you can't get the leverage for the first cut. Bolt cutters are not designed to cut case-hardened material as is found on high-security padlocks. Using two firefighters, one on each handle, is a dangerous proposition. The bolt-cutter blades may dimple, or even shatter, and the hinge mechanism of the bolt cutter itself may break apart under the tremendous pressure being applied. If you encounter case-hardened materials, select another tool!

Special Uses

Bolt cutters can be used to remove wire lath or mesh during overhaul. Sometimes this is the only method to remove such material. The bolt cutter can be used in conjunction with a hook. With the hook, knock the plaster loose. With the bolt cutters, cut the staple that holds the mesh to the wall, then peel back the loosened mesh with the hook.

The bolt cutter can also be used to twist off, not cut, battery cables on cars and trucks. Do not cut or break the posts off the battery; just twist the cables off. Breaking battery posts is hazardous because it destroys the battery casing and allows battery acid to escape.

Regarding using bolt cutters to cut energized wires: I will not discuss that in this book. Leave cutting electrical wires to power companies. Firefighters don't cut power lines.

In-House Modifications

There are no recommended in-house improvements that can be made to the bolt cutter. They come from the manufacturer ready to use, with good grips on the handles and solid cutting surfaces. Don't be cheap. Buy good cutters.

The French hitching wrap (described for the eight-pound pickhead axe) using cord and tape may improve the grip a little, but it isn't really necessary. It will look different, yes, but there is no practical advantage to be gained.

Limitations

The bolt cutter is a limited-use tool. It is designed to cut materials, and whatever materials it can cut depends on the type and quality of bolt cutter you purchase. Bolt cutters in general do not work well in tight areas. The handles must spread out far enough to allow the jaws to sufficiently clench the material. If you are working high or low, it may be difficult to get the leverage to bite down. Some materials are very tough, and you need a great deal of upper-body strength to pull the cutter closed and chomp through.

Bolt cutters are best used on light locks and chain, fencing material, shackles and hasps, and plastic-coated cable. When faced with a heavier material, it may be wise to choose another tool.

CHAPTER 2: CUTTING/STRIKING TOOLS

Don't skip this chapter! Although you may think that the tools discussed here are the same as those described in Chapter 1, they are not!

In this chapter, we will look at some basic cutting tools that have the additional capability of performing striking operations. The ability to strike another tool or an object is essential for many forcible entry procedures, ventilation techniques, and other multifunction tasks that we are required to perform on the fireground. The cutting/striking tools will be the first half of some cardinal tool sets that we will discuss in later chapters.

SIX-POUND FLATHEAD AXE

When fire departments first began buying equipment, they bought what was readily available to them. They still do. The availability of a tool doesn't make it the right tool for the job, and this particularly applies to the six-pound flathead axe. It is much too light to be an effective tool. The firefighter wielding it must do so with great strength to make the strike effective, making him dangerous to be around if he misses his mark.

Standard Uses

The six-pound flathead axe was designed for wood cutting, whether at home or in the forests of Oregon. With a razor-sharp blade, it is a great tool for felling trees. However, it has very little use on the fireground due to its light weight.

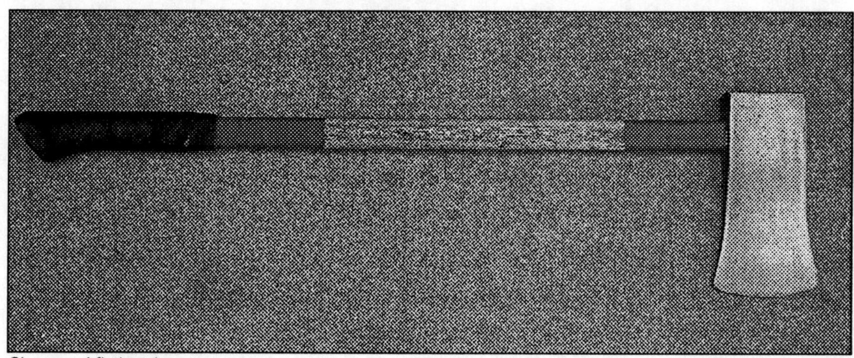

Six-pound flathead axe.

The flathead axe is primarily a cutting tool, but you also have the option of using the reverse side of the head as a striking tool. It is not a misuse of this tool to use it as a hammer, since it was intentionally designed for driving wedges into trees and logs in the lumber industry. When compared with the pickhead axe, the flathead axe is more versatile, although the fire service has been somewhat slow in recognizing that. In a lot of departments, there are far more pickhead axes than flatheads.

As with all tools, you must practice using the flathead axe to be effective. As always, the effectiveness of the tool is limited by the skill of its operator. A flathead axe is different from the lumberjack's axe. You must hold and swing it differently. The accuracy of your swing is important. The more times you can hit your target exactly where your previous swing landed, the less time you will spend completing the entire operation.

Take a proper stance. Always remember to use short strokes and to swing no higher than your shoulders. Use the weight of the tool; don't push it down. You will be ineffective if you become tired. Accuracy is key. As always, angle your strikes—a dead-on bounce will flatten your nose or, worse, break your SCBA face piece.

When you encounter a hard surface, you may need to use another tool to help make a good purchase point. Otherwise, flip the axe over and try smashing the material. Just like using the blade side, do not swing the axe over your shoulders. Let it fall accurately, and try to strike repeatedly in the same area. Weighing only six pounds, this axe will not make a dent in many materials without your having to exert great force on your swing. Don't do that. Instead, find a more suitable tool—perhaps the pickhead axe—for doing the job. Otherwise, use a halligan or other tool in combination. If the blade gets stuck, don't pull or you'll fall. Jiggle it left and right, then up and down to free it.

Special Uses

The six-pound flathead axe's special uses are really pretty limited, more so because of its limited weight than its design. It can perform several functions other than cutting, but as we will discuss in other chapters, there are other hand tools better suited for multipurpose use.

Here are some of the better uses of the six-pound axe:

- To force inward- and outward-swinging doors past their latches.
- To force double-hung windows open by prying them in their centers and pulling the lock screws out of the wood.
- To break glass.

- To make a purchase point in automobiles for hydraulic tools by prying or poking holes in the sheet metal.
- As a search tool.
- To remove moldings, baseboards, and other trim during overhaul.
- In a limited capacity, to drive/strike another tool.
- To smash ventilation holes in walls and roofs.

In-House Modifications

There isn't too much you can do for the six-pound flathead axe. It's too light to be really effective. If you must use it, see the section "Eight-Pound Pickhead Axe" to learn about modifications you can make to the handle to help it be more effective and efficient.

If you do use the six-pound flathead, grooves can be cut into the side of the tool head to make it marry together more easily with a halligan or similar forked bar.

Limitations

The six-pound flathead axe has the same limitations as its pickhead counterpart. They are:

- It's too light.
- It's ineffective as a striking tool.
- Firefighters will tire more quickly than they will using the eight-pound version.
- It's inefficient.

EIGHT-POUND FLATHEAD AXE

If ever there was a tool that should be designated the workhorse of hand tools, this would be it. The eight-pound flathead axe is an extremely versatile tool that can be used by itself or in conjunction with other hand tools to perform a multitude of tasks on the fireground.

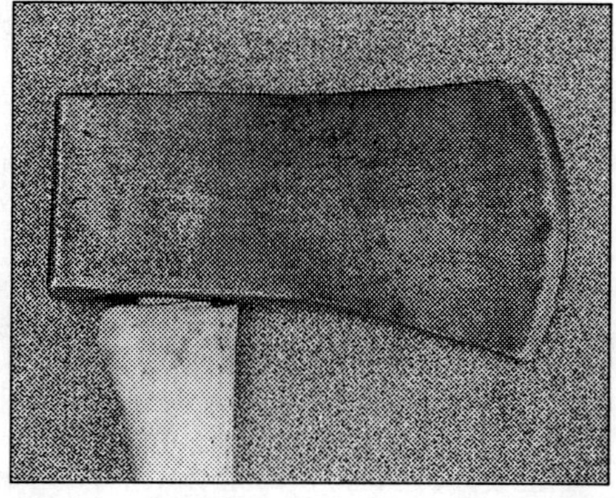
Eight-pound flathead axe.

Simply by increasing the weight of the tool by two pounds, firefighters now have a serious implement. The added weight and corresponding increase in head size give you an advantage over the six-pound flathead to cut or strike, but especially to strike.

Standard Use

Swing the axe head as you would an eight-pound pickhead. Always rely on its weight, especially when using it as a striking tool. It isn't necessary to swing it in a full arc to get its benefit as a striking tool. Hold it at waist level. Line up the flat striking surface against the tool or object that you are going to hit. Don't move that other tool! Arrange your stance so that you can strongly and effectively pivot your hips, hitting the spot you want. By putting your body weight behind it, the axe will efficiently strike and drive whatever tool you're aiming for.

The eight-pound flathead axe is an efficient roof-cutting tool.

It is especially important to practice this skill when there is no emergency. Using the axe as a striking tool is a two-firefighter operation. It must be done carefully, safely, and—most importantly—*precisely!*

Special Uses

The eight-pound flathead axe is used effectively in many forcible entry situations, overhaul operations, and rescues. It can be used at automobile fires and accidents, building collapses, and other nonfire emergencies.

In forcible entry situations, the axe is used to drive the halligan or similar tool through the door and jamb to force the latch from its keeper. It is one-half of the irons discussed in a later chapter. The axe can be used as an effective prying tool in itself but is more effective when used with a prying tool. Be very careful about any pressure you exert on the handle while prying. During most forcible entry applications, the flathead axe will be part of a tool set.

This axe can also breach walls. It can efficiently cut through them, or its opposite head can strike through them. When using the eight-pound flathead axe as a striking tool, remember your stance and energy-saving golf swing. When using the striking surface of the axe, the cutting edge is hurtling through the air on your backswing. Hitting another firefighter with the cutting edge of your axe because you didn't look before you swung will cause you to be permanently removed from that firefighter's Christmas card list. Be careful when swinging for either striking or cutting!

This tool, too, can be effective for opening walls and floors during overhaul. The amount of pressure you will need to remove building material will depend on how well it was installed, but generally you will be able to disassemble it with the axe.

In-House Modifications

See Chapter 1 for overstrike protection, handle modifications, and finger grips. Grinding two grooves into the head of an eight-pound flathead axe will marry it to a halligan more easily. Match the curve and angle of the halligan to the axe. The groove will help to prevent them from slipping apart.

Limitations

The eight-pound flathead axe is a most versatile tool. Its limitations, by now, should be apparent:

- It is heavy.
- Its primary function is to cut; striking is secondary.
- Its bigger size and weight restrict its use as a wedge during forcible entry.
- It is dangerous when used as a striking tool due to the fact that the cutting edge is moving on the backswing.

EIGHT-POUND SPLITTING MAUL

The fire service has recently discovered that the common, everyday, backyard, oak-splitting, beat-up, cheap-to-buy splitting maul is a very effective tool for many tasks on the fireground.

When was it first introduced? I don't know and couldn't find out. Many fire departments are now using this tool to replace the flathead axe in many forcible entry situations, and it is extremely popular with departments responding to newer residential structures that have strand board or simple plywood sheeting as roofing material. Its weight is effective; it provides both a cutting and striking surface, and it is inexpensive and readily available.

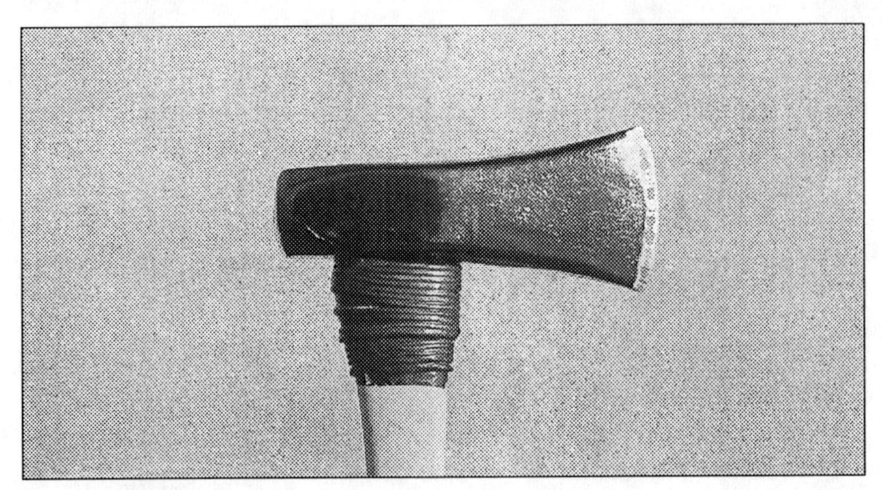

Eight-pound splitting maul.

Standard Uses

For a tool that is relatively new to the fire service, it has quite a wide range of standard uses. First and foremost, it is an excellent ventilation tool for opening up roofs. Like the flathead axe, the splitting maul will both cut and smash. Unlike the flathead, the main purpose of the maul is to destroy wood by splitting it apart, not to cut it.

Many suburban and rural departments have taken to them because they make short work of roof ventilation. Once the site for the vent hole has been selected, the firefighter smashes a hole through the roofing material rather than cuts it. The splitting maul makes short work of oriented strand board (OSB), particle board sheeting, and even plywood. No shingles have to be pulled, and because of the shape of the head, a simple twist of the handle to the right or left will free it.

Using the tool is much like using an axe. You must take a proper stance to swing it properly. Hold it in the same fashion as you would an axe. It is a little more off balance because the bulk of the weight is to the rear of the tool head. When using the cutting edge, the maul has a tendency to invert to the striking side. Hold the handle tightly and prevent it from flipping around in your hands when you swing and

Practice swinging mauls and axes both right- and left-handed.

strike. Be prepared to shift your weight around to accommodate the off-balance head. As with every cutting or striking tool we have discussed so far, accuracy is critical. Do not swing with giant strokes—no more than shoulder-high, allowing your upper hand to slide down the handle to meet your lower. Allow the weight of the tool to do the work. When it gets stuck, free it by twisting the handle to the right or left about 15 degrees. Pulling will cause you to lose your balance. Practice is critical for swinging both right- and left-handed. Always swing the cutting side of the maul so that the head strikes the surface at a slight angle. Straight strikes may bury the tool head. Remember, the maul is designed to split wood, not cut it like an axe. A heavy, deep stroke may bury it into the material being cut and make it very difficult to get out. As a striking implement, use the maul as you would an eight-pound flathead axe.

Special Uses

The eight-pound splitting maul can be used as efficiently as a flathead axe in many situations. As an overhaul tool, it is a bit cumbersome and usually by the overhaul stage of the fire you are too tired to do much swinging with this heavy tool. It is designed to destroy wood, and there are far more effective and efficient overhaul tools.

Use it in combination with a halligan for forcible entry. Although not quite the same as a flathead, the splitting maul can be part of a set of irons. The splitting maul is not an effective prying tool in itself but is more efficient when used with a prying tool.

The splitting maul is more efficient as a wall-breaching tool than the flathead axe. It can be used to cut through walls expediently or, by flipping it over and using the striking end, to smash through them. When using the splitting maul as a striking tool, the cutting surface is hurtling through the air on the backswing. Don't hit anyone! Always clear the area around you before swinging. Use your golf swing, and remember that a good understanding of building construction is essential before you breach any walls.

During overhaul, take the maul back to the rig and get a more effective overhaul tool. The splitting maul is a behemoth designed to destroy intact wood or concrete. Wood or other materials that have been burned are no match for this tool. You will also become tired. Swinging this tool to open a wall or partition that has been burned may surprise any firefighter on the other side as the head cleanly passes through the wall. Put it away during overhaul and salvage.

In-House Modifications

The modifications you can make to the eight-pound splitting maul are the same as those for the axes, outlined in Chapter 1, with a few minor exceptions.

When making the overstrike protection for the eight-pound splitting maul, make sure you use the double wrap. The head of this tool is oddly shaped (because it is designed to split wood), and overstrikes are a very common problem. Additionally, when using this tool to smash ventilation holes in roofs, the neck of the handle, just below the head, takes a real beating. Take the time to double-wire-wrap the handle at the head for overstrike protection. Granted, this is a very inexpensive tool, but snapping off the head while trying to vent a fire is dangerous (let alone embarrassing), and you don't have time to run down to the hardware store to get another. Just because it is inexpensive doesn't mean it deserves less care.

For French hitching, consider using larger-diameter rope. For some reason, this tool has more of a tendency to want to slide out of gloved hands. Increase the size of the wrap and ensure that you have a full, comfortable grip. Finger grips aren't necessary on this tool.

Limitations

The eight-pound splitting maul is an extremely versatile tool. If, however, there were such a thing as one tool that does it all, I wouldn't have written this book. Even the splitting maul has its limitations.

- It is heavy.
- Its primary function is to split wood; its secondary use is as a striking tool.
- It is unsuitable as a separate prying tool.
- When used to smash vent holes, it does not allow the use of roofing material as a shield. Smoke and fire may belch from the hole while it is being made and before the ceiling is punched. The vent hole may be limited in size due to dangerous conditions on the roof.
- It is dangerous when used as a striking tool because the cutting edge is moving where you can't see it on the backswing.
- It has limited, if any, use during overhaul.

CHAPTER 3:
PRYING TOOLS

"Give me a lever long enough and a fulcrum strong enough, and single-handedly I can move the world."

—Archimedes

Archimedes could have been thinking about firefighters when he made the above statement in the third century before Christ. His work with tools and the principle of the lever are key to the firefighters' use and understanding of prying tools. Without a basic understanding of leverage, you will not be successful in using any prying tool. All prying tools—regardless of manufacturer, style, or doodads—are levers and use that principle to accomplish a given task.

What is a lever? A lever is a bar that is free to pivot, or turn, about a fixed point. Pry bars, halligan bars, pinch bars, halligan-type bars, and many other tools used in the fire service are levers—simple levers. By applying force to one end (effort) and having a strong, fixed point to pivot against (fulcrum), a mechanical advantage is gained over whatever object (resistance) you are trying to move. The proper use of leverage is the basic principle behind using fire department prying tools. This chapter will examine several different types of prying tools and their uses. Remember that no tool will function properly if the operator fails to apply the principle of leverage correctly. When leverage is improperly used, the tool will usually slip and the object will remain quite unmoved.

Prying tools may be needed on the roof.

PRY BAR

Standard Uses

Almost every piece of fire apparatus I have ever seen carries at least one pry bar. These tools are the ultimate in simplicity because they are true levers. There are two types of bars: the pinch bar and the wedge-point bar, the difference between them being the point on the tool.

Pinch bars have only one beveled side. These bars range in size, dimension, and weight. The smaller size is usually three feet long by one inch square, weighing six pounds. The sizes of the bars continue to increase, reaching a massive size of 5½ feet long, 1½ inches square, and a weight of about 26 pounds! As can be seen in the photo below, the pinch bar is flat on the bottom and has a sloping, chisel-like bevel.

Pry bars.

The wedge-point bar has a bevel on both sides of the bar, forming a wedge point, hence its name. These bars also come in various lengths and weights, although not as varied as the pinch bar. The wedge-point bar ranges in size from three feet long, one inch square, and a weight of six pounds to five feet long, 1¼ inches square, and a weight of 18 pounds.

Both bars are limited-use tools because the fire service has invented many others that will do the work more efficiently. However, these tools should not be removed from service! By taking a little time to learn how to use them, a firefighter will be better equipped to handle many different situations. These bars are the ultimate levers, and they are inexpensive. Their value in collapse situations, heavy rescue, and even in some applications during confined-space rescues in some industrial settings cannot be matched by any other tool. A pinch bar and a wedge-point bar should be carried on at least one of your rigs. They won't get much use, but when you need them they will work!

These tools can be used in conventional forcible entry to open doors, windows, and so on. However, they should not be your first choice for forcible entry. The wedge-point bar has a slight advantage over the pinch bar in conventional forcible entry. When combined with a striking tool, the wedge point of the bar can be driven into a door or window frame. The wedge shape allows the tool to slip into the recessed areas readily, its spreading force being applied in both directions. Once the tool has been driven deeply into the door or window frame, force can be applied, and the door or window will open. There are better tools to use for forcible entry, but pry bars will work.

The advantages that these tools have over all others are their length and narrow profile. They can be used, with a proper fulcrum, to raise collapsed material, machine parts, automobiles, trees, and other fallen debris. Pry bars operate in the mud, snow, rain, smoke, and all the other environments that firefighters must work in.

Both pinch and wedge-point pry bars are the best tools when heavy objects must be lifted or moved. Their length and size adds to the amount of force they can convey. Using a pry bar when stabilizing a car with cribbing will make work easier. By using the cribbing as a fulcrum, firefighters can gently move or lift a vehicle to slip additional cribbing underneath the frame.

During overhaul operations in plaster and lath fire buildings, firefighters can insert the pry bar into the bay of the wall and make fast work of opening it. The entire length of the tool is then used, and there is no chance of snapping it as there is when doing the same procedure with an axe handle. Pinch bars are excellent tools for prying up wooden floors. By inserting the tool into the seam of the floor, bevel side up, the tool can be driven in, lifting the first piece of flooring. Then, using the tool as designed, leverage can be applied and the flooring will come up. Two firefighters, each armed with a pry bar, will make fast work of removing many floor surfaces. Once the first piece is removed, the rest will follow easily.

Baseboards, moldings, and portal frames are also easily removed with a pry bar. Often you will not even have to bend down to insert it. Simply allow the pry bar to slide along the face of the wall. Nine times out of 10, both wedge-point bars and pinch bars will find the joint between the wall and the baseboard. Pry outward, and the baseboard or molding will pop off.

Pry bars are relatively inexpensive and can be purchased at a local hardware or farm supply store. Their availability and low cost make them the preferred choice as tools that may have to be sacrificed to a fire. Pry bars can be used as chocks to hold doors open during firefighting operations. They are especially valuable for heavy-duty doors, doors with self-closures, and roll-down doors.

There is no real trick to their standard

For difficult prying jobs, use a pry bar.

use. Find a good purchase point on the object to be moved, and locate the fulcrum as far down the shaft as possible. Both of these bars are basic levers. If you understand leverage, you will use these tools effectively.

Special Uses

The pry bar has a variety of special uses, none of which have to do with using the tool as it was designed!

Besides being used as a door chock as already mentioned, the pry bar has several other unconventional uses on the fireground. Technically, these abuse the tool; however, such techniques work, and in the long run no significant damage is done to it.

The first special use of the tool is as a securing post for ground monitors or deck guns. The pry bar can be driven into the ground like a fence post and the monitor secured to the pry bar to prevent it from walking away. The pry bar must be driven in deep and may be difficult to remove after the fire is out, but it does provide a secure post for lashing a gun.

Pry bars can also be used for securing ladders to windows. Place a long pry bar horizontally across the inside of a window and tie a ladder rung or beam to it using a rope. Apply sufficient tension on the rope to pull the pry bar up tight against the interior window framing or walls. Note: The pry bar should be substantially wider than the window!

Pry bars also make good handles for carrying basket stretchers or other heavy objects. Properly lashed, the bar provides handles for several firefighters to grab at once.

In heavy search and rescue or collapse situations, pry bars can be used to move very heavy objects, such as machinery. They can also be used to break up concrete and rocks. In trench rescue, they can serve as digging bars.

In-House Modifications

There are no specific in-house modifications that should be made either to the pinch bar or the wedge-point bar. The bevels of both tools should be kept crisp and clean. Using a pry bar with wet hands is a problem because the bars are steel and therefore get cold and slippery. Wrap the handles with friction tape or hockey stick tape. Also, bicycle hand grips (remove the streamers) can be added to the handle end to enhance your grip.

Some departments have added opposing handles so that two

firefighters can manipulate the bar while standing opposite each other.

Limitations

When used properly, the lever is a fantastic tool. In bygone days, railroad men known as gandy dancers would use pry bars in a rhythmic motion to straighten sections of rail, ties, and ballast. With a proper fulcrum, these simple levers may be the solution to the fireground problem you are facing. Don't overlook these tools. Their limitations include:

- They are heavy.
- They have limited uses.
- Due to their length, they are difficult to use in tight spaces.
- They take up compartment space.
- They are best used in pairs.

DETROIT DOOR OPENER

Standard Uses

From the Motor City comes this venerable fire service tool that saw wide use in departments worldwide. Today, this tool is usually relegated to a running board bracket on a reserve truck, and most firefighters have no idea what it is or how it is used. It has been (and should be) replaced by smaller, more powerful hydraulic tools or more efficiently designed pry bars. This is an ignominious end for a veteran fire tool.

The Detroit door opener is a curious contraption that is actually a lever system capable of delivering tremendous force when used correctly. The opener provides its own fulcrum and is adjustable to different positions to deliver maximum force. Although an old tool, the Detroit door opener still functions well when used correctly and will have devastating effects on today's standard locks and door closures.

The tool consists of several simple parts. The first and main part of the tool is the lever part. It looks like a pry bar (which it basically is) with a narrow handle on one end and a wide point at the other. The wider, slightly thicker curved end is the fulcrum of the tool. Force will be applied to this part of the tool.

Attached to the fulcrum end is a pivoting joint that connects the extending bar to the main part of the tool. This extending bar will receive the effort from the main bar and transfer it. Force will be transferred to the third part of the tool: the clawed swivel pad at the end of the bar. The swivel pad is attached to the portion of the extending

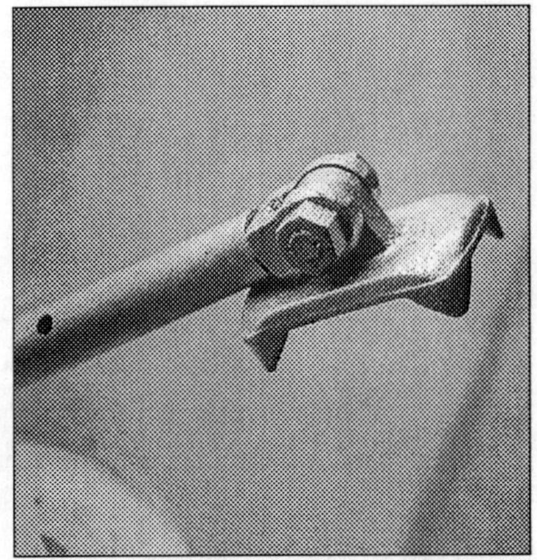

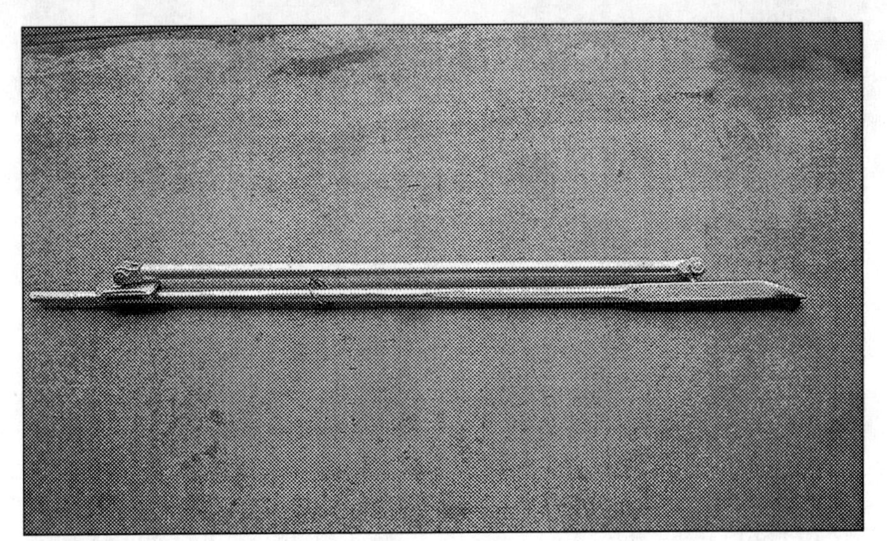

Detroit door opener.

bar that pulls out after the adjusting pin is removed. The bar is extendable to allow for different types of door assemblies and to allow the firefighter to set up the tool to achieve maximum force. The overall length of a standard Detroit door opener, when folded, is about 50 inches. Fully retracted, the extension arm is 39 inches, and it extends outward to approximately 66 inches. The tool is made of forged steel and weighs 23 pounds.

The Detroit door opener will require an operator with good upper-body strength when it is used on steel doors set in steel frames; doors that have high-security locks, padlocks, or hasps; or doors held closed with magnetic locks. The fact is, the Detroit door opener will probably not work in these situations no matter how strong the operator is. The tool was originally designed before security became a real issue in the United States. The Detroit door opener is usually unstoppable in standard residential situations, apartments, and hotels/motels. The key is understanding the lock mechanism you face, the type of door, and the use of the tool.

The Detroit door opener, when used properly, exerts extreme force on a locked door.

Working with the Detroit door opener is simple—there are two methods that you can use. The first method is recommended if you are not familiar with the tool or are poor at judging distances by eye. First, size up the door you are going to force. Ensure that it is not a high-security door held closed with sophisticated locks or high-security devices.

Place the fully closed tool vertically against the door, with the fulcrum end resting on the doorsill or floor. Use a piece of chalk (You do carry a piece of sidewalk chalk in your pocket for marking searched rooms, don't you?) or your hand to mark the position on the tool that corresponds to the height of the lock on the door. If there are two locks, mark whatever position on the tool falls between them.

Next, lay the tool down on the floor in front of the door, perpendicular to it. If you didn't mark the tool with chalk, maintain your hand position. Now, mark the spot on the floor. Repeated training with this tool will make you a good judge of what distance is needed.

Slide the tool toward you, stand it up, and place the fulcrum on the marked spot. Place the clawed footpad directly on the lock or between the locks. A quick jerk on the lever handle will help set the footpad into the surface of the door. Once the tool is set, apply force on the lever handle toward the door. Some firefighters recommend a quick backward motion, but the overwhelming majority of firefighters who are familiar with this tool recommend a slow, even, increasing pressure. A fast motion may throw you off balance, the tool may slip from your hand, or some other malfunction may cause you to be injured. Slow and steady does it. This tool is a true lever, and the force you are exerting on the door lock assembly is tremendous. Continue to apply more force on the main lever handle until the door lock gives in to the pressure.

The second method of using the tool is simply to judge the distance out from the door, set the fulcrum in position, and place the footpad on the door, making the adjustments required. This method is a little quicker, but handling the tool can be clumsy. With practice, this will probably be your preferred method.

Although not readily available on the commercial market today, the Detroit door opener remains in the inventory of many departments. It should not be the primary choice for forcible entry situations, but it shouldn't be overlooked, either. The Detroit door opener will do the job, and it is much faster to use this tool than it is to wait for another set of irons or some other forcible entry tool set. In an emergency situation, knowledge of how to use the opener may mean the difference between getting into the fire building and a delay and worsening of the situation.

Special Uses

The Detroit door opener has no special uses. It is an extremely limited-use tool, designed only to force open doors.

In-House Modifications

The only modifications that should be made to this tool are sharpening the fulcrum point and wrapping the handle for a better grip. The Detroit door opener was engineered for a specific task. Any modification to the tool may be hazardous to your health.

One exception to that is to keep the fulcrum point of the tool well dressed and sharpened (not razor sharp) to facilitate its bite into the floor during initial setup. The tool must not be driven into the floor, since the fulcrum must pivot as force is applied.

Another exception would be to add a wrapping of friction tape, hockey stick tape, or possibly a smaller French wrapping similar to that added to the handles of the cutting tools. The steel bar can become very slippery, and the addition of tape or a French wrap will help toward maintaining control of the handle.

A neat trick with the tool is to mark estimated door sizes on the handle—that is, take the tool to various buildings in your area and premeasure it against residential doors, commercial doors, interior doors, and so on. This will save you a step in the initial use of the tool and is more accurate than using chalk or your finger to mark the distance.

Limitations

As with any tool, practice will make you more proficient. The Detroit door opener has one use: It is solely designed to force open lightly or moderately secured, inward-swinging doors. That's all. Its other limitations include:

- It is heavy.
- If you're not careful, it will pinch you.
- It is clumsy and hard to carry.
- It may not work against high-security locks.
- It won't work well on steel doors in steel frames.
- It requires a lot of practice to master.
- The firefighter using the tool needs good upper-body strength.
- Setting up the tool is time-consuming.
- Using it in the dark or in smoke-filled areas is dangerous.
- It's difficult to maintain control of the door if the tool is being used by a lone firefighter.
- It has been replaced by hydraulic tools and efficiently designed pry bars.

CLAW TOOL

Standard Uses

This veteran fire service tool has several names. Usually the first name of the tool is the city that either claims its development or was a heavy user of it; hence, the city name became synonymous with the tool.

Commonly, the claw tool is called the L.A. claw tool, the Hayward claw tool, or the New York claw tool. No matter who developed it or where it was developed, the claw tool is a functional and durable item

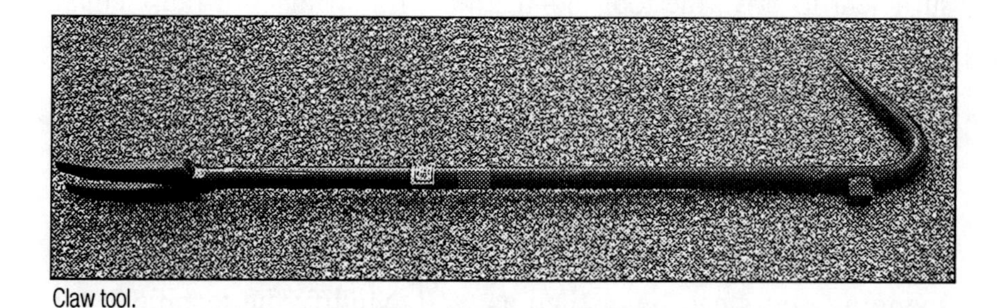

Claw tool.

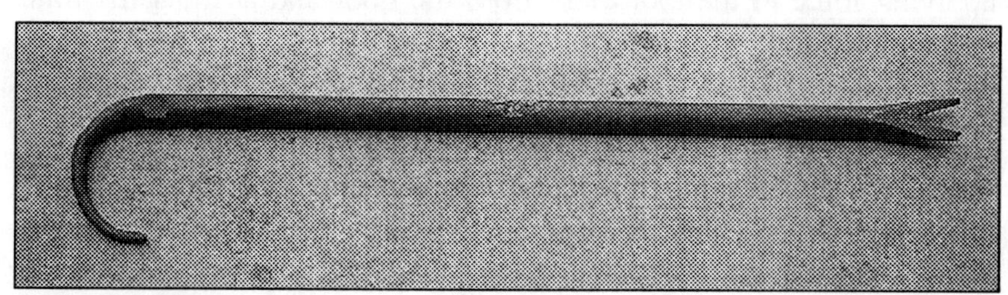

Original-type claw tool.

for use in the fire service. There are other prying tools that are superior to it, but as with all the other implements in this book, you should learn to use what you have!

The standard claw tool is approximately 42¾ inches long and weighs around 14¼ pounds. There are some shorter versions and a couple that are slightly longer. On one end of the tool is a slightly curved, beveled fork. On the opposite end is a hook, much like a shepherd's crook, which tapers down to a sharp point. Just about where the hook curve begins to form, there is a square knot of steel, which is to be used either as a striking surface or as a fulcrum when prying.

The claw tool is a step up from the previous prying tools we have discussed. It is a multipurpose prying tool and, although not extremely versatile, it can perform a great many tasks on the fireground when used correctly.

Forcible entry procedures can be performed easily with the claw tool. The biggest disadvantage is that it has no striking surface at the end of the hook that might be used to drive the fork into place. The striking surface is on the opposite side of the hook and is used for driving the hook.

When combined with a striking tool, the claw tool can be driven in and conventional forcible entry techniques used. The fork can be placed between the door and frame to pry the lock bolt out of its keeper on inward-swinging doors. To do this, place the fork of the claw tool about six inches above or below the lock with the bevel side against the door and the claw tool slightly angled toward the floor or ceiling. Strike the claw tool with the back of a flathead axe, driving the forked end past the interior doorjamb. Be extremely careful! The claw tool does not have an engineered striking surface. The blows should be sufficient to drive the tool in—neither wild swings nor muscleman swings! Easy does it. As you strike the tool, slowly move it perpendicular to the door to prevent the fork from penetrating the interior doorjamb. Exert pressure on the claw toward the door, forcing it open. You must practice this technique. The hook end of the tool does not present a large target for striking, and the hook point is pointing toward the door and the firefighter holding the tool. Outward-swinging doors can be opened as well. Modify the technique by applying force to the tool away from the door and at the same time prying outward to get the lock bolt out of its catch so that the door will open toward you.

The tool's length makes it a great lever, but at more than 42 inches long it is a tight fit in narrow hallways and rooms. This tool is sharp at both ends! When using the fork, pay attention to the hook end. If the claw tool is not well set and happens to slip, the point of the hook end can rip open skin, tear off face pieces, and generally do lots of nasty damage.

The claw tool can also be used to open windows. The beveled ends of the fork, if they are well dressed and maintained, can easily be slid between the bottom rail of the window and the sill. Keeping in mind the principle of leverage, set the fork well. Once you slide the fork in as far as you can get it, pry down on the bar. The screws holding the window lock should pull out.

You can also use the fork end to shut off residential and light-commercial and industrial gas valves. The distance between the tines of the fork is sufficient to allow a good bite on the gas valve.

You can twist off padlocks by sliding both sides of the shackle into the fork and twisting or prying downward with a sharp motion. Proper size-up is key during this, however. If the hasp or other device that the lock is attached to is not of a good, strong material, *do not* twist off the lock! The hasp metal will twist, and although the shackles will break off, the metal of the hasp or other device may be so twisted that the door still won't open.

Flip the claw tool over and use the hook end for breaking padlocks. This tool makes short work of most of them, and the striking surface of the tool can be used, making this technique much safer than twisting.

Insert the hook point of the tool in the shackles of the lock. Slide the hook down until it is firmly wedged. Strike the claw tool sharply with the back of a flathead axe or sledgehammer. Continue to strike the tool, driving the hook in. The shackles will break and the lock will fall off as you force the widening steel of the claw tool through the shackles. The curve of the hook end allows you to slip in the claw tool behind the entire hasp assembly. Applying force will tear the lock, staple, and hasp right off the building or door. This works well for wood framing, but do a good size-up of the lock assembly before trying it on steel framing or masonry. You may make more of a mess and screw up the hasp assembly so much that you'll need a power saw to get in.

The claw tool makes an excellent overhaul tool as well as a forcible entry tool. Here its length is an advantage, since the firefighter doesn't have to bend so much when using it to remove baseboards and flooring. The hook works well for prying up flooring and subflooring, and the striking knot will act as an effective fulcrum when you rock the tool back to apply force to the floorboards. You can also slide the fork under the floorboards and easily pull up the flooring. The 42-inch length is a great mechanical advantage for heavy-duty work like overhaul. The claw tool is not a tool you want to work with over your head for very long: Fourteen pounds is a lot of weight to hold aloft to remove upper window casings and door trim. The tool is best when used at waist level or below.

Special Uses

The claw tool has some special uses, but they are limited. The hook end is an effective ripper for getting a firm bite into ductwork and other light metal surfaces. When inserted, the knot of steel that had been used as a striking surface becomes a great fulcrum, and the ductwork or metal can easily be removed.

The curve of the hook end is a great handle when using the claw tool to open walls. Once a hole is created in the wall, slide the tool, forked end down, into the hole. Grasp the hook end and pull toward you, and the entire length of the tool will be pulled through the plaster and lath or wallboard.

Some tool manufacturers have added two lugs on one side of the hook. These lugs can be used to assist the firefighter in opening standpipe valves. They can be inserted into the spokes of the standpipe handwheel. The claw tool then becomes a cheater bar for the firefighter to open the stuck valve. Take extra caution, however. Some of these handwheels may be pot metal rather than steel (despite what the codes say) and will shatter under force. Use the tool with finesse to coax the valve open. Make sure you are turning it in the right direction. Extreme force may break the wheel, or the tool may slip and you'll fall flat on your face. Take a balanced stance, and apply slow, even pressure to the tool. Have a pair of channel locks or vise grips with you just in case.

During vehicle extrication, the hook end can be driven into the sheet metal and used to peel it back. This is especially useful when trying to get into the hood of a burning car.

In-House Modifications

One modification that can be made to improve the grip on the claw tool handle is to make a French hitching wrap of it, between the fork and the striking knot, as described on pages 11-12. The wrapping should start about two inches up from where the fork joins the bar and end about the same distance below the striking knot. As mentioned earlier, lugs can be added to the curve of the fork for opening standpipe valves.

Limitations

The claw tool is a multipurpose tool. When used correctly, it is an effective forcible entry and overhaul tool. It does, however, have limitations, which include:

- It's generally too long (42¾ inches).
- It's heavy (14¼ pounds).
- It's sharp at both ends.
- The striking surfaces available are limited.
- It's not as efficient as a halligan bar.

KELLY TOOL

Standard Uses

The kelly tool was a difficult tool to research—the fire service has about five to eight different implements that are called kelly tools. Some are standard construction pry bars, modified flat bars, crowbars, and other types of tools. In this section, we will look at the real kelly tool, which is a forerunner of the modern halligan bar. Its invention is credited to Captain John F. Kelly of Hook & Ladder 163, FDNY.

The real kelly tool is a steel tool, approximately 27 inches long and weighing about 12¾ pounds. The ones I have seen have all been painted. On one end of the tool is an adze-type head. The adze is flat and about two inches wide. The edge of it is beveled, and the top of it can function as a striking surface.

On the other end of the tool is a large, heavy-duty chisel. The blade is about three inches wide, slightly tapered, and sharp. It is set at a 90-degree angle to the adze end. The chisel has no curve to it. At the top of the chisel where it joins the bar is a heavy steel collar or ring, which prevents the tool from being driven into a surface too far. The kelly tool is 15¾ inches shorter than the claw tool, which makes it nice for tight hallways, but at 27 inches it's just a little short to be a real effective lever.

Properly used, the kelly tool is a very effective forcible entry instrument. This is the first tool we have looked at in this section that has an effective adze end. The adze end can be used, in combination with a striking tool, for a quick and effective means of forcible entry.

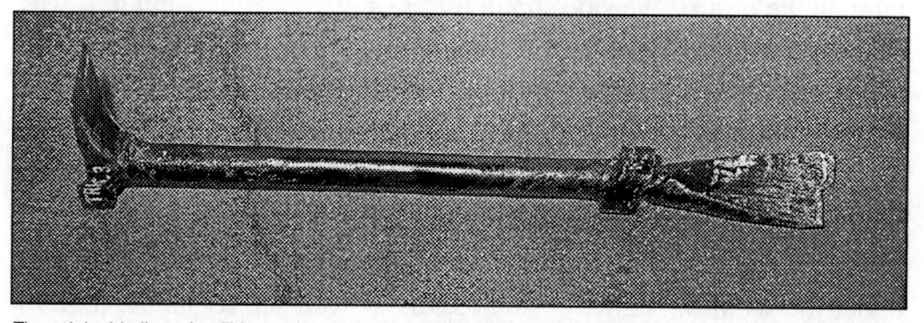

The original kelly tool, still in service at the University of Illinois Fire Department. Note the relationship of the adze end to the chisel end.

Perform a proper size-up of the door. If it must be forced, take the kelly tool and a striking tool. Slide the adze of the kelly tool in between the door and jamb about six inches above or below the lock. If there are two locks, place the adze between the locks (for inward-swinging doors). With the striking tool, drive the adze into the doorjamb. Strike the back of the adze (*do not* strike the chisel end). The kelly tool adze has no curve to it, so you are going to drive it into the door frame. That's the way this tool works: The adze will split the door frame. Drive the adze deep, and ensure that you have a good fulcrum. Once you set the tool, you can apply pressure in two ways. Either push on the bar toward the door, splitting the door frame and forcing the door open, or push down on the bar and twist the adze from the flat position to its wide position. By twisting the adze in this manner, the door will be spread away from the door frame the width of the adze. That is generally enough to get the door to open. You may have to apply quite a bit of force to make the tool work. Be careful. Make sure you have good footing and, most importantly, maintain control of the door. Balance yourself so that, when the door does open, you don't follow through and end up in the room. Not real healthy if the room is full of fire.

The chisel end of the tool can also be used to force doors. It is most effective on inward-swinging doors but will work on outward-swinging ones as well. The steel collar of the tool does get in the way, and the fact that the chisel end has absolutely no curve to it makes using it a messy operation. Without a doubt, you are going to split the door frame.

To use the chisel end, place the chisel about six inches above or below the lock, the kelly tool slightly angled toward the floor or ceiling. Drive it as you would a claw tool. As before, try not to penetrate the interior doorjamb. This will be tough, but do your best. It makes a difference. Exert pressure on the kelly tool toward the door, forcing it open. Chances are pretty good that the chisel end will be driven into the interior doorjamb and split it. Oh well, the door will be open and you can continue with the task. Just because this tool is a little on the destructive side shouldn't prevent you from using it. It was designed as a forcible entry tool, and it's a good one.

You can also use the kelly tool to open windows just like the claw tool. Keep the bevel of the adze end and the chisel end well dressed and maintain the bevel. Insert the chisel end of the tool between the bottom rail of the window and the sill. Prying on the bar should pull the screws out of the window lock. You can also slip the adze end under the window frame. Once the adze is well set, rotate the tool 90 degrees in either direction. The window lock will pop.

Special Uses

The kelly tool is a multipurpose forcible entry tool. Because of its design, it has no real special purposes other than functioning as a pry bar. This doesn't mean there aren't any special uses—just that there aren't any that the kelly tool performs exceptionally well.

In-House Modifications

The kelly tool can be modified to become a more modern type of tool. The problem is that, if you don't have access to a good machine shop, there isn't too much you can do to it. Some of the modifications may actually reduce its efficiency or weaken it.

One modification that can be made is to machine an A tool into the adze end of the bar. This will increase its functionality by making it suitable for both conventional and through-the-lock forcible entry. It is recommended that you take the tool to a professional machine shop to have this done. Bevels, edges, and curves are all important issues when making an effective A tool. Also, you don't want to weaken the adze end. By removing steel to cut the A tool, the adze may become too weak to depend on during heavy prying situations.

On the other end of the tool, a V can be cut into the chisel to be used as a gas shutoff, a nail puller, or whatever else you have the greatest need for. Just as with the adze end, the tool may be weakened.

A chain link can be welded onto it as a snap hook attachment for easy hoisting. The link can also be used to attach a snap hook and utility rope to allow firefighters to vent windows from the roofs of buildings, wrecking ball-style.

The grip on the kelly tool can be improved by adding friction tape and French hitching as described earlier.

Limitations

The kelly tool is an excellent tool compared with the standard pry bar or the claw tool. It does have several limitations, however, including:

- It's too short for good leverage.
- It's heavy (12¾ pounds).
- There is no curve to the adze end.
- The chisel end is not curved or forked.
- There is no point or hook on the adze end.
- The tool has been replaced by the halligan bar.

SAN FRANCISCO BAR

Standard Uses

The San Francisco bar was developed and originally manufactured by the San Francisco Fire Department shops. Modeled after the kelly tool, the original implement was welded together from heavy steel and weighed about 16 to 18 pounds, depending on its length. The biggest complaints about the original tool were its length and weight.

The San Francisco bar of today is similar to the original bar but has more of the characteristics of the Chicago patrol bar and the halligan bar.

The modern San Francisco bar is of forged steel, with an adze on one end and a curved fork at the other. The adze end of the tool is at a 90-degree angle to the fork. There is no pike or pick point on the tool. At 30 inches long, it is a powerful lever.

A unique feature of the tool is at the fork end. On the back side of it, at the top, where the hexagonal bar joins the fork, there is an added half-rounded piece of steel. This half round gives the fork a built-in fulcrum and makes this tool very effective in forcible entry and overhaul situations. The tool, depending on the manufacturer, also comes with an added ring at the fork to attach to a utility rope for hoisting or dropping over the edge of a roof to ventilate upper-floor windows. The fork is approximately six inches long and has a gentle curve. Each fork has a beveled edge, with the bevel on the top side (concave side) of it.

The adze end is at 90 degrees to the fork and does not have a point or pick. It, too, is gently curved and has a flared-out beveled adze that widens from the tool head to the adze end. The beveled edge is on the bottom side of the adze. The bar itself is of $^{15}/_{16}$-inch hexagonal steel and weighs approximately nine pounds.

The reengineering of this bar from its original version has produced a lighter, better-balanced tool as compared with the original. It is an outstanding multiuse pry bar for conventional forcible entry and standard overhaul techniques.

For forcible entry, there are several ways to use the San Francisco bar. To open an inward-swinging door, place the fork about six inches above or below the lock with the bevel side against the door, the tool slightly canted as you would a claw tool. Strike the top surface of the adze, in line with the bar as much as possible. Use a flathead axe or other striking tool to drive the forked end of the San Francisco bar past the interior doorjamb. Swing the striking tool just hard enough to drive the tool in. Use caution—the adze end of the tool is at 90 degrees to the fork. The bar man must be aware of the location of the sharp end of the adze so that he doesn't walk into it or have the bar slip and hit him.

Exert pressure on the San Francisco bar toward the door, forcing it open. The adze end can also be used very effectively to open inward-swinging doors. Place the adze end of the San Francisco bar six inches above or below the lock. Drive the adze into place with a striking tool. Make sure the adze has a good, deep purchase in the door frame. Pry downward with the forked end of the San Francisco bar.

For doors that open outward, place the concave side of the fork toward the door and cant the tool slightly toward the floor or ceiling to get a better purchase. Drive the tool using a flathead axe or other striking tool. As the tool is driven, move it perpendicular to the door to prevent driving the fork into the jamb. When the tool has spread the door as far as possible, force the adze end of the San Francisco bar away from the door.

Using the adze method for outward-swinging doors is also effective with the San Francisco bar. Place the adze of the tool six inches above or below the lock. Drive the adze into the space between the door and the jamb. When you are sure the adze is sufficiently into the space, pry down and outward with the fork end of the San Francisco bar.

The fork end of the San Francisco bar can used to open windows. The offset adze end of the tool makes a great handle, allowing you to move out far on the bar, increasing your leverage. The adze end can also be used for prying. Once the adze is well set, rotate the tool 90 degrees in either direction. This will pop the window lock.

The San Francisco bar can be used for breaking padlocks. This tool makes short work of most standard padlocks, but the twisting technique must be used since the tool has no pick or point to drive through the shackles to shatter the lock. Size up the lock that must be removed. If there is no way to attack the attachment itself (that is, pry off the hasp or the entire assembly), the fork can be slipped over both sides of the lock shackle to twist it off. This method can only be used on locks attached to heavy-duty staples or locking assemblies. Lightweight metal may twist, and even when the lock shackle breaks, the remaining metal may be so twisted that you won't be able to open the door. The adze may be forced behind the hasp on wood-frame doors. A severely damaged lock assembly will probably require the use of a power saw or hydraulic tool to get through.

The San Francisco bar makes an excellent overhaul tool as well as a forcible entry tool. Either the adze end or fork end can remove baseboards, casings, moldings, and flooring. The added half round of steel on the fork end is outstanding for prying up flooring and subflooring. The half round is a very effective fulcrum when you rock the tool backward to apply force to the floorboards. You can also slide the adze under the floorboards and easily pull up the flooring.

As with all prying tools, do not try to lever more than the tool can handle. The tool will bend if it is used on extremely heavy materials or materials that are well attached or supported. Granted, it will take tremendous force to bend this tool, but it can happen and has happened. Don't overtax it.

Special Uses

In its modern configuration, the San Francisco bar is an excellent pry bar. The half round at the fork end gives this implement a little bit more versatility than a straight pry bar or other similar tool. Because it was originally designed by firefighters for fireground use, the San Francisco bar is extremely effective for forcible entry and exit, ventilation, overhaul, and rescue.

There are no true outstanding special uses that are unique to the San Francisco bar. It is an effective, well-balanced pry bar and a great tool if you carry one on your apparatus.

In-House Modifications

You might have an A tool machined into the adze end, as you would to a kelly tool. Welding a chain link to it as a rope attachment is also helpful.

Limitations

The San Francisco bar is a very efficient and effective prying tool—if it had been anything else, it would have disappeared into fire department tool history by now. Like any other tool, though, the bar does have its limitations. These include:

- It's very heavy (original bar only).
- It lacks a pick point or hook.
- It's too long (original bar only).
- It's sharp at both ends.
- The offset adze end can get in the way when using the fork end.
- It is ineffective for through-the-lock entry (without modification).
- It is not as efficient as a halligan bar.

CHICAGO PATROL BAR

Standard Uses

The Chicago patrol bar was developed and used by the Chicago insurance patrols. The original bars were manufactured by a company whose name has been lost to history. The bars found their way into the Chicago Fire Department from the patrol and can still be found on some Chicago fire apparatus. This is a well-built, massive tool. The biggest firefighter complaints about it were its original length and weight.

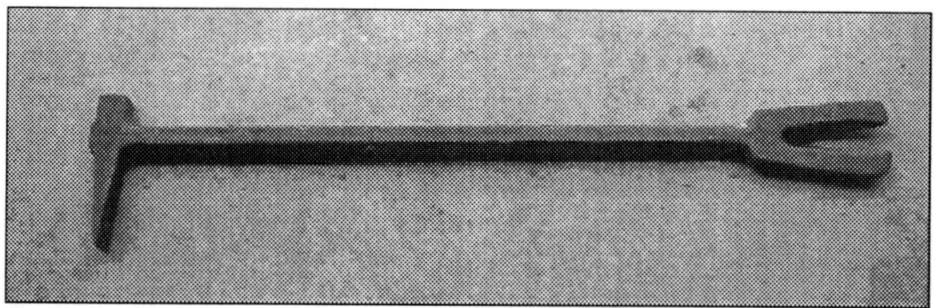

Chicago patrol bar.

The true Chicago bar is actually a form of the claw tool. The original bar had a forked end and a curved hook tapered to a point. Today's Chicago patrol bar is actually a modified halligan-style item that has maintained the original patrol bar's hammerhead configuration on the adze end of the tool.

The modern Chicago patrol bar is of forged steel, with a hammerhead adze on one end and a curved fork on the other. The adze end of the tool is at a 90-degree angle to the fork, although I found a modern patrol bar on one Chicago Fire Department truck company that didn't have the offset tool ends. There is no pike or pick point on the tool. The bar is 30 to 35 inches long and weighs nine to 11 pounds.

The patrol bar, depending on the manufacturer, may be designed with an added ring at the fork to attach a utility rope for hoisting or dropping over the edge of a roof to ventilate upper-floor windows.

The fork is approximately six inches long and has a gentle curve to it. Each fork tine has a beveled edge, with the bevel on the top side (concave side) of it.

The original manufacturer of the Chicago patrol bar has been lost to history, but the tool lives on in the Windy City.

The adze end resembles that of the San Francisco model, and the bar is also of 15⁄16-inch hex steel. The hammerhead on the adze was designed to allow the patrol bar to be used as a striking tool as well as a pry bar. The original idea was to have a firefighter carry two patrol bars rather than a more limited-use flathead axe or sledgehammer. One bar would be the pry bar; the other would be the striking tool. Firefighters would then have two pry bars for extreme rescue situations requiring heavy prying or a prying and striking tool.

The modern bar is better balanced than the original and much lighter. It is an outstanding multiuse pry bar for conventional forcible entry and standard overhaul techniques. With its redesign, which combined the best of the original patrol bar and the halligan bar, the advantage of having a hammerhead was lost. The tool is too light to be an effective striking tool—nine pounds overall. With its improved prying capabilities, the mass required for easily driving another tool has been traded off.

For forcible entry, there are several techniques for using the Chicago patrol bar. To open an inward-swinging door, drive the forked end past the interior doorjamb. Try to keep the striking tool level with your hips, and swing using your hips rather than just your arms. The same cautions and methods that apply to the adze end of the San Francisco bar also apply here. Outward-swinging doors are opened in virtually the same manner as with the San Francisco bar. Refer also to the earlier sections on the claw tool and San Francisco bar regarding opening windows, breaking padlocks, levering, and effecting forcible entry.

Special Uses

Because the tool was originally designed by the Chicago Fire Department for use in situations common to the Windy City, it is effective for all phases of firefighting. The tool has enough strength built into it that it can be used during heavy rescue operations.

In talking to many Chicago firefighters, I have found that they like the tool and that both the modern and original versions are retained in Chicago firehouses. The tool comes up just a little short in some situations, making it somewhat of a limited-use tool when compared with others.

In-House Modifications

Machine an A tool into the adze, and weld on a chain link. See modifications relevant to the kelly tool, page 37.

Limitations

In general, this is an efficient and effective prying tool. Both the original, shop-made version and the modern design are still in use. Most firefighters opt for the halligan-style tool, but some make good use of the patrol bar and love it. Closely resembling the San Francisco bar, it does have its limitations. These include:

- It's very heavy (original bar only).
- It lacks a pick point or hook.
- It's too long (original bar only).
- It's sharp at both ends.
- The offset adze end can get in the way when using the fork end.
- It is ineffective for through-the-lock entry (without modification).
- Although it has a built-in hammerhead, it is too light to be an effective striking tool.

HALLIGAN BAR

Standard Uses

If you're wondering why the halligan bar is the very last prying tool in this chapter, the answer is because the halligan tool is the newest of all the firefighter's prying tools to be put in service!

The true halligan bar was developed by Deputy Chief Hugh Halligan of FDNY during the 1940s. Chief Halligan had seen several other tools used by the fire department, both alone and in conjunction with other tools. Combining the very best features of the claw tool and the kelly

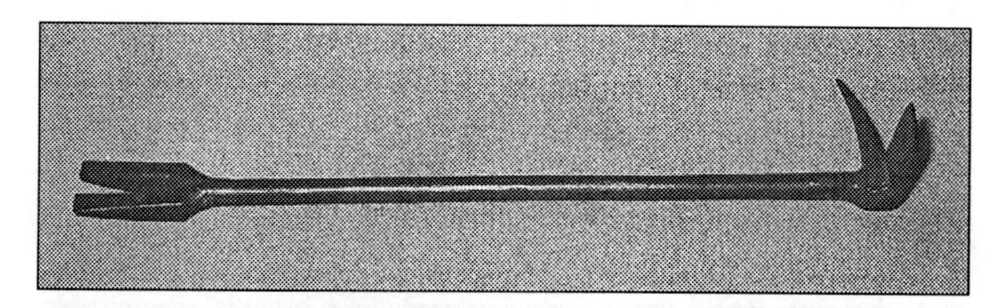

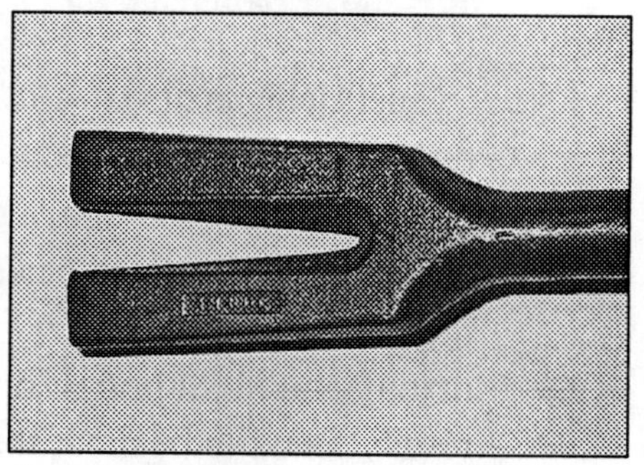

The original halligan bar. Chief Halligan's signature is on the top tine of the fork pictured. These halligans were hand-made.

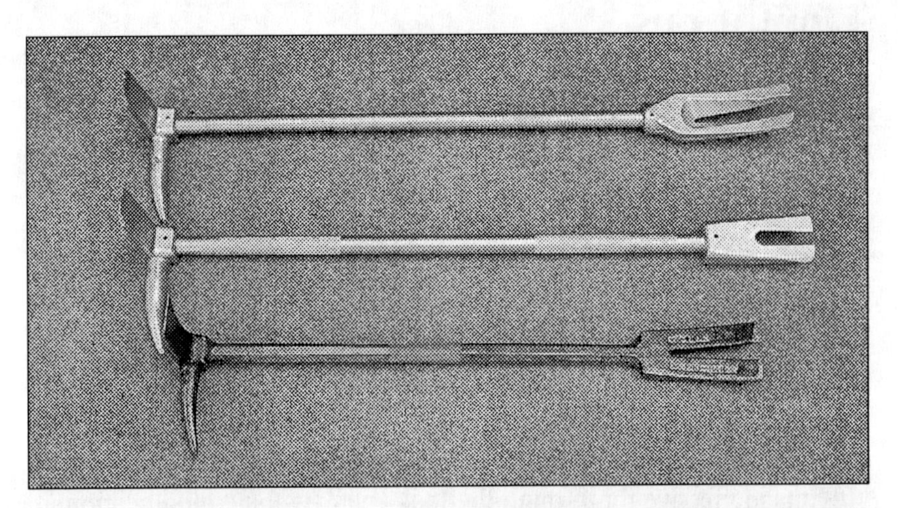

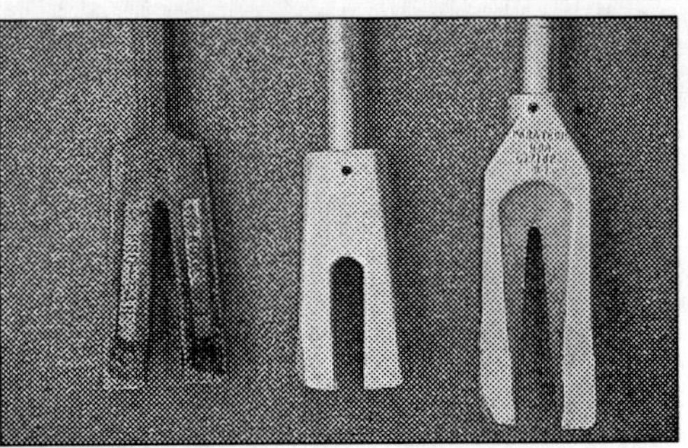

Halligan-style bars come in many varieties. Some types are more efficient than others.

tool, Chief Halligan designed and developed his own implement. The bar was much lighter (8½ pounds) and shorter than the claw tool yet longer than the kelly tool; it had a sweeping adze and a redesigned fork. Chief Halligan also added the pick point or hook to his tool, a remnant of the curved hook found on the claw tool.

After a lot of research and testing, the original halligan bars were made, each one requiring more than 18 man-hours to produce. Chief Halligan's signature appeared on every one.

Throughout the late 1960s and early 1970s, when large U.S. cities were under heavy fire duty, the halligan bar was the mainstay forcible entry tool, used not only by FDNY, but also by the Boston, Philadelphia, and other large urban departments that needed a substantial, dependable tool.

Chief Halligan died in 1987 at the age of 92. His patent has long since run out on his forcible entry tool. Its continued use by the fire service has led a variety of tool manufacturers to produce "halligan" bars. Some are sweet; some should be used as boat anchors. This section will cover the best and most-used version, based on its design and use, not its manufacturer. Although pry bars based on the halligan design are readily found throughout the fire service, some are inferior to the real item in quality and design. They may not produce the same effect as a true halligan bar, and may be difficult to use in many forcible entry situations. When making your purchasing decision, consider the curvature of the tool ends, the thickness of the metal, and whether or not the item is forged.

There is a lot of confusion about the difference between the different brands of halligan bars. Why aren't they all called halligans? The answer is money. Patent rights, royalties, lawyers in three-piece suits—all of that. These issues are a long way from the fireground where we need to know how to use the tool in our hand to force open the door or force our way out. We'll let somebody else worry about the money.

The uses of the halligan tool have been obtained from line firefighters. This section of the chapter will describe applications preferred by firefighters who use the tool a lot and who taught me how to use it.

Working with the halligan bar is still a matter of leverage, just like every other prying tool discussed so far. The design of it allows for multiple functions being contained in one tool, but leverage is the key. Halligans are available from 20 inches to 42 inches long. Its use will dictate what length you need. A 30-inch halligan is the best bar for use on a day-to-day basis for conventional forcible entry, ventilation, overhaul, and other standard fireground activities. Shorter bars have their place, as do longer bars, but optimize your tool selection with a 30-inch version. Weight is also a consideration, and the 30-inch bar weighs only nine pounds.

The tool should be a single piece of forged steel. Tools that have heads welded or pinned on don't function as well as the forged item. The tool has areas susceptible to failure should it have to be placed under extreme stress, or if it has been poorly maintained and becomes corroded. Tubular shafts are also substantially weaker than a solid piece of steel.

The adze end should gently curve and flare out slightly from the tool shaft to the

The halligan bar is efficient at forcing inward-swinging doors.

tip. The adze should be beveled, with the bevel on the bottom side of the adze. At a 90-degree angle to the adze is a hook point or pick. Wide at the base where it joins the tool, the pick should also taper and curve, ending in a relatively sharp point. The head of the tool also has two striking surfaces. The top of the tool (on top of the adze) and the side opposite the pick are both striking surfaces that have been designed to receive heavy blows from a striking tool.

The shaft of the tool is usually at least $^{15}/_{16}$-inch hexagonal steel. The hexagonal shape adds strength and rigidity, and the many faces improve your grip when holding the tool.

The fork is broad and tapered. It should be at a minimum six inches long and taper into two well-beveled tines. The bevels are located on top of the tines. The spacing between the tines allows gas valves, padlock hasps, and other objects to be levered by the tool. The bottom side of the fork is called the beveled side, and the top, dished side is the concave side. There are no striking surfaces on the fork. Depending on the manufacturer, the tool may or may not have a ring attached just above the fork for snapping on a utility rope for hoisting and ventilation purposes.

The standard 30-inch halligan will perform numerous functions when used properly. It is an excellent bar to use in conventional forcible entry.

For inward-swinging doors, two firefighters can make quick work of even a well-secured door. The irons man (the guy holding the bar) places the fork of the halligan about six inches above or below the lock. The bevel side of the fork should be against the door. If it is a steel or metal-clad door, you may need to flip the bar over and put the concave side against the door. Angle the bar slightly toward the floor or ceiling. The second firefighter strikes the halligan with the back of a flathead axe when the irons man tells him to. His intention is to drive the forked end of the halligan past the interior doorjamb, not to Timbuktu. As the bar is struck, the irons man slowly moves the bar perpendicular to the door being forced to prevent the fork from penetrating the interior doorjamb. The irons man will probably really have to lean on the bar to get it to move after each strike.

The halligan will also work efficiently on outward-swinging doors.

Once you are sure that the tool is sufficiently set, and the halligan is really fully perpendicular to the door, apply pressure on the bar toward the door, forcing it open. Make sure you can maintain control of the door.

Another method for inward-swinging doors is to drive the hook of the halligan completely into the doorjamb six inches above or below the lock. Push the halligan bar down and the door will open. It is very important to push down. In trying all the different methods, I discovered that you work twice as hard physically when you pry upward, and the door doesn't usually open! Use leverage to your advantage, especially in this case. Pushing down to lever the door open is much easier than pulling up! Pulling up may not open it at all.

Outward-swinging doors are another story. Flush-fitting doors can be forced using either the adze end or the fork end of the halligan. The fork-end technique is very effective. Drive the tool in the same manner as discussed above. The adze end is similiarly useful for outward-swinging doors. Drive the adze deep. The bar should stand out by itself perpendicular to the door when you're finished driving it in. When you are sure the adze is sufficiently into the space, pry down and out with the fork end of the halligan.

Another possible method to open an outward-swinging door is to insert the adze end of the bar underneath the door. Pull back on the halligan. You may be able to get a good purchase point and enough leverage to open the door with this technique.

If you run across recessed doors or those with a wall next to the lock side of the door, use the adze end of the tool. Place the adze end of the halligan six inches above or below the lock. Drive the adze into place. Pry downward and outward with the forked end of the halligan. This operation is a bit clumsy. Be extremely careful when driving the tool into place. One firefighter can do it, but not too gracefully. Two firefighters are needed; don't hurt each other. Work as a team, and always remember that the irons man calls for the strikes.

An advantage of the halligan bar is its ability to be used as a lever to remove burglar bars and screens from windows. This is a new technique for the halligan bar. It requires that you have a small length of chain and a heavy-duty hook. The hook is attached to the burglar bar, close to the attachment point. The halligan is placed up against the building in a three-legged stance. Use the adze end of the tool, and create a triangle up against the building with the adze and pick. Drop one of the chain links over the pick so that the chain is as tight as possible. Pry down on the end of the halligan. The force should pull the attachment device, screw, or bolt out of the wall and loosen the burglar bar. A cable setup for this purpose is also available from a fire tool manufacturer. The halligan must be modified for the cable setup. The modification is a simple notch machined into the tool.

The halligan will easily remove child-safety window gates. To perform this, strike the vertical supports of the child gate where it attaches to the window frame. Use the adze of the halligan to remove the screws from the window frame. Pry upward if possible to pull the screws out and down.

During roof operations, the pick or hook of the halligan can be driven into and through the roofing material to determine whether fire is below you at that location. If you do find fire, don't stand at that spot too long.

The halligan bar also makes an excellent safety step on roofs. Hold the halligan so that the pick is against the roof surface. Drive the pick into the roof with a striking tool. The halligan is now a step to use when operating power saws or swinging an axe on a roof.

The tool can also be used as a step-up for parapet walls or other areas that are just too high for a normal step, especially when you are wearing bunker pants. Set the tool down fork-end first. Turn the adze end toward the wall or strong surface. Tap the fork down into the roof or floor. Then tap the halligan head, and set the corner of the adze and the point of the pick into the wall material. You can step on the tool to get up to the next level. Reach down and recover the tool.

The halligan bar also makes an excellent ladder brace. The pick or hook of the tool can be driven into a porch roof or even the ground and used to foot the ladder if a second firefighter is unavailable to do so. The tool can also be used to level a ground ladder. Lay it on the ground so that the adze and pick or hook form an inverted V. Level the ladder on the tool.

During venting or rescue operations, firefighters often encounter chain-link fences. The pick or hook of the halligan can be used to break the wire attachments that hold the chain link to the fence posts. Be careful! When you release the fencing, it will roll up and may take you with it. As you break the wires, hang on to the fencing and assist it as it rolls up.

The bar can also be used to ventilate upper-floor windows from the roof. Some tools are manufactured with a small ring on the shaft, just above the fork. Attach a utility rope to the ring. Lower the halligan over the edge of the roof. Lean over and look. Center the tool in the window below. Mark the utility rope. Hoist the tool back to the roof. Holding on to the marked spot on the utility rope, toss the halligan off the roof. Arc it out so that it swings back and breaks the window below. For thermopane windows, you may have to toss it more than once.

The halligan is a great tool to use during overhaul. This tool will do

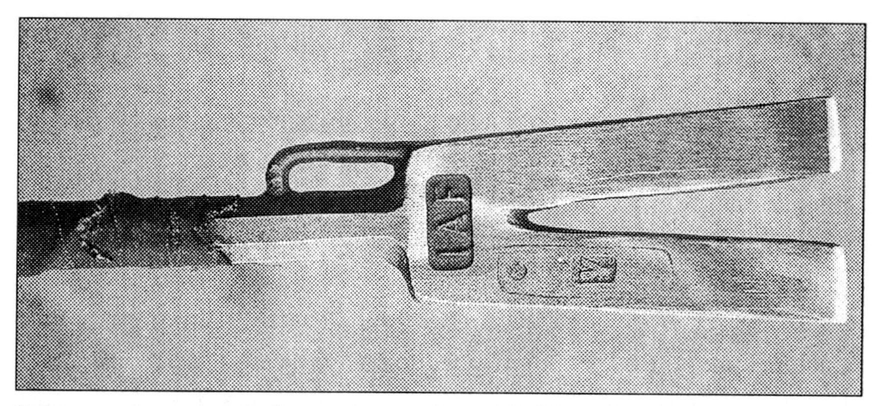

Halligans can be purchased with snap rings, or a large chain link can be welded onto the shaft, just above the fork and opposite the pick.

ten times more work than a closet hook. It is capable of pulling trim, molding, baseboard, framing, gypsum board, and plaster and lath. Together with a striking tool, a firefighter will be able to dismantle an entire room.

The halligan bar is also good at opening windows and padlocks. The easiest way to open a padlock with this tool is to use the pick like a duck-billed lock breaker. Insert the pick of the bar through the shackles. Using a striking tool, drive the pick deep into the shackles until the lock breaks. This technique may not work on high-security padlocks, but it does work on most standard types.

Special Uses

The fork of the halligan can be used to "cut" the glass out of automobiles. First, pierce the glass with the halligan; then, grip the edge of the glass with the fork. Rocking the tool up and down in a scissorlike motion will cut the laminated glass from the car. This same technique will allow the tool to cut some light sheet metal. The adze end of the halligan is an extremely versatile tool. It can be used very efficiently as a bolt-head cutter when married up with a sufficiently weighted striking tool. Bolt heads can be sheared off and the object can be removed. This is a very effective and efficient way to remove wire mesh screens from schools and factory windows. The pick or hook on a halligan bar also makes it an excellent tool for lifting manhole covers. By inserting the pick, then lifting up on the tool, the manhole cover can be levered off or back into place. A last special technique that the halligan bar can be used for is self-defense. Burglar alarms, fences, and razor wire are not the only security systems that people use. Guard dogs are a threat to your life. They don't recognize you as a firefighters—only as intruders. The halligan bar will give you the reach and deadly force needed to kill a threatening animal.

In-House Modifications

If your tool was not manufactured with a snap ring on it, a heavy-duty chain link can be welded on. Weld the link just above the fork on the side opposite the pick. By installing it opposite the pick, the tool will dangle with the pick pointed toward the window when using it to ventilate windows from the roof.

The tool can be made even more versatile by having an A tool machined into the adze end. This technique will cause you to lose quite a bit of adze mass and may ruin the adze for cutting bolt heads. In most cases, however, the overall effect is an improvement in the tool rather than a detriment.

See the section "Eight-Pound Pickhead Axe" for instructions on French hitching a halligan bar.

Limitations

There are few, if any, limitations to a true halligan bar.

CHAPTER 4: STRIKING TOOLS

This chapter was going to be the quickest and easiest chapter to research and write ... or so I thought. What is a striking tool? What do you call it? If it can be used to strike another tool, does it fit into this section?

Nationwide, the fire service has terms and terminology that are either universally used in the business or are specific to one department. A prime example of this is for striking tools. Is it a maul or a sledgehammer? What is the difference?

Going to the dictionary was no help at all. *Webster's New World College Dictionary* defines "maul" as coming from the Latin *malleus,* meaning hammer. It describes it to be a "very heavy hammer or mallet, often of wood, for driving stakes, wedges, etc." The dictionary defines "sledgehammer" as "a long, heavy hammer, usually held with both hands."

We're still nowhere toward defining what these striking tools are and which is which. Quite a debate raged and, finally, a conclusion was reached. Based on modern industrial definitions and descriptions, the difference between the tools described in this book is as follows:

Sledgehammer—A heavy, long-handled hammer with striking surfaces located on opposing sides of the tool head.

Maul—A heavy, long-handled hammer with a striking surface on one side of the tool head and another type of surface, such as a splitting wedge or pick, on the other side.

Using these definitions, this chapter will be dedicated to sledgehammers. Other striking tools can be found in Chapter 2, Cutting/Striking Tools.

While the sledgehammer is the mainstay striking tool for many trades other than just the fire service, it is used for one standard purpose and one purpose alone: to strike an object or another tool.

Okay, so it seems to be too simplistic, but it is an important and true statement. There are very few tools available to the firefighter that are designed to strike other tools. The flathead axe and the eight-pound splitting maul are the only two that come to mind. Remember, most of the tools discussed (with the exception of cutting tools) are designed to be striking surfaces, not tools we use to strike with. As mentioned in Chapter 3, the San Francisco bar, Chicago patrol bar, and halligan bar can be used as striking tools, but they are not as efficient as a tool that was so designed. What else do you have available to you that will be capable of performing the task of driving a halligan bar into a doorjamb or knocking a steel high-security door off its hinges? If your department performs collapse rescue, what fire service tool is available to drive stakes and large nails to build cribbing and shoring?

The sledgehammer is the thing to have available when a heavy hammer is needed. They're broken down into two major categories (long-handled and short-handled sledgehammers), then into weight and use classifications.

LONG-HANDLED SLEDGEHAMMERS

Standard Uses

The first category of sledgehammer is the standard hardware or builder supply-type of hammer. The standard model has a 36-inch handle made of pine, hickory, ash, or fiberglass. It is also available with a 32-inch handle. Stay with the 36-inch handle for the standard sledgehammer handle.

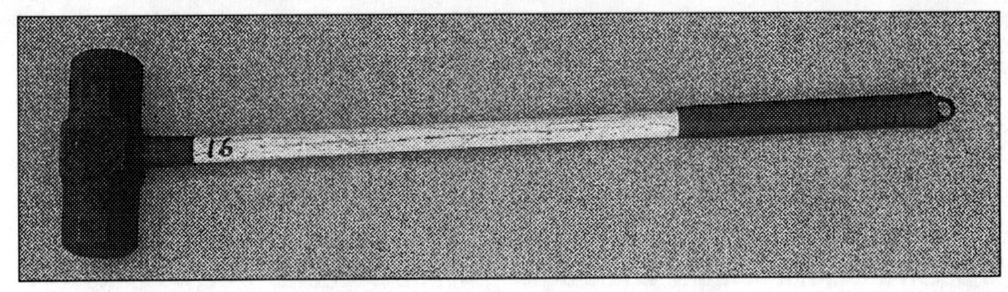

16-pound long-handled sledgehammer.

Physics plays a role here also. When using sledgehammers, we need to be aware of two of Sir Isaac Newton's Laws. Newton's second law of motion states that Force = Acceleration × Mass. This shows how much force (push or pull) there is on an object. The amount of force depends on how much mass the sledgehammer has and its acceleration as we swing it.

Newton's third law of motion predicts how objects will interact. The law may be stated either in terms of action and reaction or in terms of opposing forces: For every force there is an equal and opposite force. For every action there is an equal and opposite reaction.

This means that when a force is put on an object, a reaction will occur of the same strength but in a manner opposite that of the force. What all this boils down to is this: When you select a sledgehammer to use as a striking tool, the weight of the tool is important. For you as a firefighter to accomplish a given task, like driving a stake into the ground to hold a deluge gun in place or striking a halligan bar to break a lock, Newton's Laws come into play, and they determine how hard you have to swing to get the job accomplished. If you have to swing too hard to accomplish the job, get a heavier hammer!

Weight is the key. In the fire service, sledgehammers should be in the weight range of eight pounds, 10 pounds, 12 pounds, and 16 pounds. No more or less than that. The mass of the tool determines how effective it is, as well as your ability to swing and strike effectively and accurately. The old adage "Let the tool do the work" is important here.

Eight-pound sledgehammer—This weight category is the most effective for using the long-handled sledgehammer for forcible entry techniques such as striking pry bars, duck-billed lock breakers, hammerheaded picks, and other similar tools. At eight pounds, there is enough mass, and you will be able to swing it with enough controlled velocity to be extremely effective. Anything less than eight pounds must be swung too hard to be effective.

Ten-pound sledgehammer—This weight is an intermediate size. Forcible entry techniques involving striking the pry bars, duck-billed lock breakers, hammerheaded picks, and other items are still possible, but it is quite a bit more difficult to do them accurately. A miss with an eight-pound sledgehammer can do great bodily harm, and a 10-pound sledge will definitely ruin your day and spoil your good looks.

The 10-pound sledgehammer will quickly and effectively open doors by knocking them off their hinges. It will shatter locks and breach walls. It is more difficult to swing accurately, and the firefighter swinging it will tire very quickly, especially under heat and smoke conditions or while wearing full protective clothing.

Twelve-pound sledgehammer—The 12-pound sledgehammer is the first weight that should be reserved for situations requiring heavy force. It is too heavy to be used as a tool to strike another tool in forcible entry. The mass of this tool and the tremendous blows it will deliver with little effort make it ideal for breaching walls, breaking concrete, driving stakes, and other heavy jobs. The 12-pound hammer is an excellent tool to have available, but it will be a very limited-use tool. It is extremely difficult for the average firefighter to swing accurately without tremendous amounts of practice.

Sixteen-pound sledgehammer—This is a tool of mass destruction. The mass of this tool makes it very difficult to swing, impossible to swing accurately without practice, and a very tiring experience. This is the very best tool to use when something blocking your way must be destroyed. In researching this tool, I used a 16-pounder on a locked door in an old motel complex that was being demolished. In four moderate swings, I knocked the entire door assembly out of the wall—door, frame, everything fell into the room. It was impressive but impractical. The idea had been to open the door. If this had been real, and there were a victim just inside, he or she would have been crushed by a 125-pound steel door assembly.

A special note must be made about swinging a long-handled sledgehammer, regardless of its weight. As a firefighter, there is no need for you to swing a sledgehammer over your head and out of sight. Control of the tool is paramount for your safety and the safety of others.

Newton's Laws will help you if you keep them in mind. If you find yourself swinging the tool into a position out of your peripheral vision field, change to a heavier sledgehammer. There should never be an instance where you will lose sight of the tool due to your swing. Not over-the-head or way-around-sidearm, off behind you. Keep the tool head in sight, and maintain good hand-eye coordination to deliver the maximum effective blows on target. Missed blows mean just that many more strikes you'll have to make to accomplish the job.

Special Uses

There are no special uses for the long-handled sledgehammer. It is a hammer and should only be used as such.

In-House Modifications

It is hard for even the handiest firefighter to come up with modifications for it. The sledge is designed to be a striking tool. The striking surface should be clean and free of any burrs, flaking metal, or rust. Use sandpaper or a wire wheel to remove any rust or flaked metal. Its striking surface should be smooth, and its head should be oiled to forestall rust. Add overstrike protection. French hitching is also useful. These modifications are both described in the section "Eight-Pound Pickhead Axes," Chapter 1.

Limitations

The very same design that makes the long-handled sledgehammer an effective tool is its biggest drawback: weight. Sledgehammers are heavy. They need to be heavy to be effective, but you don't want to have to tote them around throughout the fire.

Their other limitations include:

- Long handles make sledgehammers difficult to use in tight areas such as hallways and closets.
- Sledgehammers are limited-use tools; they function solely as striking instruments.
- When used improperly, sledgehammers are dangerous.
- In certain instances and uses, sledgehammers create flying debris.

CUSTOM SHORT-HANDLED SLEDGEHAMMERS

Standard Uses

This tool is not available from any manufacturer (yet). A popular diversion in the Chicago Fire Department and many of the surrounding departments is to make a custom short-handled sledgehammer from a long-handled one. I believe this idea also came from FDNY. I can't verify that, but it doesn't really matter, either.

To make a custom or cut-down sledgehammer, you need either an eight- or 10-pound, long-handled sledgehammer with a wooden handle. *Do not use a tool with a fiberglass handle!* The eight-pound sledge is preferred.

The idea is to customize the length of the tool, fitted to your size. Measure the tool and cut the wooden handle to make the entire tool length, head included, between 30 and 33 inches. After making the cut, sand the end of the tool handle smooth.

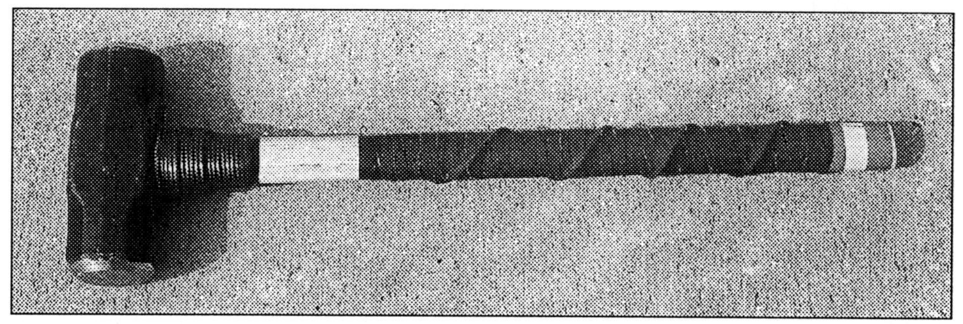

A custom short-handled sledgehammer.

Wrap the handle in French hitching as described earlier. Add overstrike protection and an extra wrapping at the handle end to form a knob that will prevent the tool from slipping out of your hands. You now have a custom short-handled eight-pound sledgehammer.

The short-handled sledgehammer will do all the functions of the long-handled sledgehammer, especially in tight areas such as hallways and stairwells, although with less velocity.

Newton's Laws cannot be forgotten. By cutting down the handle, you will not be able to swing the short-handled sledgehammer the same as you would a long-handled one. You are giving up some velocity for more accurate swings and more control. That's fine, but don't try to use the short-handled sledgehammer in instances where a long-handled tool is needed.

Special Uses

The short-handled sledgehammer has a variety of uses that would be considered special when compared with the long-handled one. The short-handled version can be carried with a halligan or other pry bar. This modified set of irons can be used in areas of high security—on steel doors and buildings where it may be simpler to breach the wall rather than force open the door. Its reduced length makes it an easier tool to use in two-man forcible entry situations, and it is much safer than using a flathead axe in most cases, since there is no cutting surface to dodge.

The custom short-handled sledgehammers are a matter of personal preference. They are inexpensive to make, and they don't take up much room on the apparatus. You can never have too many tools.

In-House Modifications

The whole tool is an in-house modification!

Limitations

As with most tools, the custom short-handled sledgehammer has its limitations. These include:

- It functions only as a striking tool; there is no cutting surface.
- Firefighters may have a tendency to swing it too hard to make up for the shortened handle and consequent loss of velocity.

CHAPTER 5:
POLES

This chapter will cover all of the different tools mounted on poles that are used in the fire service today. Like the pickhead axe, the pike pole is actually a very ancient tool that has its origins as a military weapon. Pikes were long poles with metal spearheads at their tips. The spearheads weren't arrow-shaped as you might think but instead were long and narrow and often triangular in shape. Entire armies were made of pikemen.

Various navies also used them—you have probably heard of the term boarding pike. It was basically the same tool used by armies, only shortened to fit aboard ship. Sailors would use these pole arms to repel boarders or to make boarding raids or would use them in defense onshore.

Along with the pikes, navies also had hooks. Boat hooks were very common and are still used today. The boat hook retrieves material dropped overboard, captures small boats as they move alongside, and keeps your bass boat from leaving the dock without you.

Like many other tools discarded by the military, the fire service found a tremendous need for these tools. Early hook and ladder companies, so called because they only carried hooks and ladders, would arrive at the scene of a fire and proceed to tear down the two adjoining structures with their hooks to prevent the spread of fire.

Early colonial firefighting consisted of whatever tools were readily available. Because the colonists had arrived here aboard ship, the only firefighting tools on hand were naval boarding axes, pikes, and hooks.

This chapter is organized a little differently from the other chapters in the book. To save your eyes from reading the same material over and over again for each of the poles described, the standard uses of pike poles will be discussed in general, independent of the style of tool head. Then, each type of pike pole tool head will be separately discussed, outlining their special uses and limitations.

PIKE POLES

Pole Materials

The pike pole (just the pole, not the tool head) is available to the fire service in three basic materials: wood, metal, and fiberglass. These materials are standard fare when ordering pike poles. The shape of the pole can be round, oval, or I-beam. Wooden handles can be pine, hickory, or ash; fiberglass handles can be solid round or oval, hollow core, or solid I-beam; and metal handles can be stainless steel, plain steel, or aircraft steel.

Pike pole lengths vary from three feet up to 16 feet and longer. The types of poles outlined in this book are the six-foot, eight-foot, 10-foot, 12-foot, 14-foot, and 16-foot pike pole.

The material your poles are made of, their shape, and their length are really dictated by your department. As a general rule, all pike poles are very serviceable or can be made serviceable. The biggest problem encountered with poles is their diameter. There are some pike poles that have diameters of 1½ inches or larger, made of fiberglass. These poles are extremely difficult to work with because they don't fit in your hand, are very slippery when wet, and are a pain to stow on the apparatus.

Length is somewhat important also. So-called closet hooks look neat on the apparatus and are easy to carry but really serve little function on the fireground. To use a closet hook that is anywhere from 2½ feet long to four feet long usually means you have to be in the closet to use it, and you end up pulling all the stuff down on top of you.

Short poles—that is, shorter than six feet—also require you to work with your hands up over your shoulders to pull trim, open ceilings, and so on. This is very tiring, and you will not be as effective. I've seen firefighters jumping up and down to sink their three-foot closet hooks into the ceiling of a standard residential bedroom. If the idea had been to conserve your energy by carrying a lightweight tool, does it make any sense to have to work three times as hard to get the tool to function properly? Nah! That's silly! A halligan bar is a far superior tool to have with you; it will outperform a closet hook every time.

Pole Selection by Size

Pole length should be dependent on the type of structure you are working in. Just as you would select the proper size wrench to fix the plumbing on a sink, you need to select the proper length pole for the type of structure you are working in.

Six-foot pole—Consider making the fire service standard six-foot pole the minimum length of pole carried for basic firefighting operations. A six-foot pole will easily reach most ceilings, trim, windows, and other parts of a residential structure that you will need to get to. It can be carried into and out of the structure safely and

Pike poles allow you to vent windows from a distance.

maneuvered easily inside. A six-foot minimum length will allow you to work with the pole out in front of you, so that materials will drop out and away from you rather than on your head. It also allows you to work with your arms in a low position, which is less tiring.

Eight-foot pole—The eight-foot pole is even better than the six-footer for allowing you to work with the tool out in front of you rather than to stand in the drop zone of overhead material. That extra two feet also allows you to work with your arms at about waist level, and you will be able to work more effectively and be less fatigued. An eight-foot pole can also be maneuvered in standard residential structures, although a bit more care should be taken. The eight-footer should be the minimum-length pole taken into a commercial structure or light industrial building.

Ten-foot pole—Okay, this pole is not suitable for the standard residential home ... or is it? Do you have any of those new, 5,000-square-foot homes in your jurisdiction? Large homes often have atriums, high cathedral ceilings, balconies, and many other building features that may make the use of a 10-foot pole mandatory.

This pole is a bit cumbersome in standard residences, or it just plain won't fit. You need to have one available, though, not only for overhauling large houses and commercial structures but also for ventilation operations. The 10-foot pole would be the most useful pole for you in commercial buildings and light industrial buildings.

An additional problem begins to show up with poles starting at the 10-foot length and longer: The pole material becomes a factor. Some poles become extremely flexible at 10 feet—most notably the oval fiberglass one. It's like trying to work with a fishing pole! The tool head end bends and wobbles and is a real pain. For 10-foot poles or longer, consider solid wood or round fiberglass. The best handle for long poles is probably the fiberglass I-beam, which gives you the most strength with the least amount of flex. Such poles are heavier but are more solid and easier to work with.

Twelve-foot pole—The 12-footer is too long a pole to use in a residence. There may be special circumstances where it is needed in a standard residence, but those will probably be rare. The 12-foot pole should, however, be the minimum-length pole you take to the roof of any structure—residential, commercial, industrial, or otherwise. The 12-foot pole will allow you to reach into your vent hole to push down ceilings and stay out of the smoke and possible roaring fire that may push out of the hole.

This length is probably the standard pole to take with you in malls, department stores, and other large commercial and industrial structures. Most stores have at least 12-foot ceilings, and they may be as high as 14 or 20 feet. Know your jurisdiction and compare that with the lengths of poles available to you.

Sixteen-foot pole—This length pole is usually used in large commercial structures and industrial facilities. It is very heavy and cumbersome and takes practice to use. Although the use of the 16-footer is seldom called for, make sure that you have some available, especially for commercial and industrial occupancies.

Adjustable-length poles—These poles are weaker than solid poles and thus less effective, no matter what tool head is on the working end. Adjustable poles aren't worth it.

Handles on Poles

Many manufacturers provide (and fire departments can add) different types of handles or knobs on the end of the pole opposite the working tool head. Some of these handles, listed below, can be used as tools.

D handles—D handles are available on many poles. They make the most sense on six- and eight-foot poles, but you can get them on any length pole or go to the hardware store and add your own.

The addition of a D handle makes the tool somewhat easier to use. When using the tool on the roof to push down ceilings, insert the D handle into the vent hole. It will spread the force to a larger area when you push the pole down, opening up a greater amount of material each time you apply the pole. The D handle on the pole should be rounded, with no edges that will catch on things in the hole. The opening in the handle should be large enough to accommodate a gloved hand. When using the tool to open walls or ceilings, use the tool head—your hand should not be inserted into the handle during this operation. The D handle should be used only when pulling on the tool. Inserting your hand into the handle and then driving the working end of the tool into a ceiling, wall, or floor is a good way to break your wrist, pull ligaments, or otherwise tear yourself up. Do *not* use the D handle when *pushing* with the pole!

Although helpful on the downstroke, a D handle on a pike pole should never be used to push or to strike.

Under certain conditions, D handles can be detrimental. Your pole is not as versatile because the smaller, rounded end of the pole is not available to make initial holes for purchase points. D-handled tools are also harder to stow on the rig.

Gas shutoffs can be added to poles.

D handles have become popular for pole handles.

Ram knobs—Some poles have a metal ram knob on the bottom. This little knob can be used as a mini-battering ram on shorter poles. It is effective for opening some doors but is most efficient when used to start holes in walls or ceilings. The knob end is also useful in determining where the end of the pole is so that your hands don't slip off when pulling or pushing.

Gas shutoff—Slots are cut into the end of the pole to turn gas valves in residential or commercial buildings. It's a nice feature, because you're at least six feet away when you do it, rather than on top of the meter with standard gas wrenches or spanner wrenches with shutoff slots.

Prying ends—One type of pole that will be discussed later in this chapter has a prying end for opening vents, hatches, doors, and so on. The tool is a roofman's hook and was designed by FDNY for use by the roofman in his operations. The prying end is broad and curved. It can easily be driven into areas to get a good purchase to lever open a variety of materials and fixtures. Pike poles are not levers, however. The prying end of the roofman's hook is part of an all-metal hook. The roofman's hook was designed for prying. The standard pike pole wasn't.

Double-headed poles—Poles with working heads at both ends are available. What may seem to be a great idea—two poles at once—isn't really practical.

Fire service pike poles are direct descendants of military arms. Most often there have been no modifications to the tool design itself. The pole doesn't know it's not in the military anymore. It is an instrument of destruction and not something that you as a firefighter want to have close to your body when it is being used.

By putting a tool at each end of the pole, both ends become dangerous. At the end where you are standing and working, you are pushing or pulling a sharp, pointed tool. If the tool slips, if you fall, or if a variety of other mishaps occur, the pole will revert to its ancestry and tear your guts out.

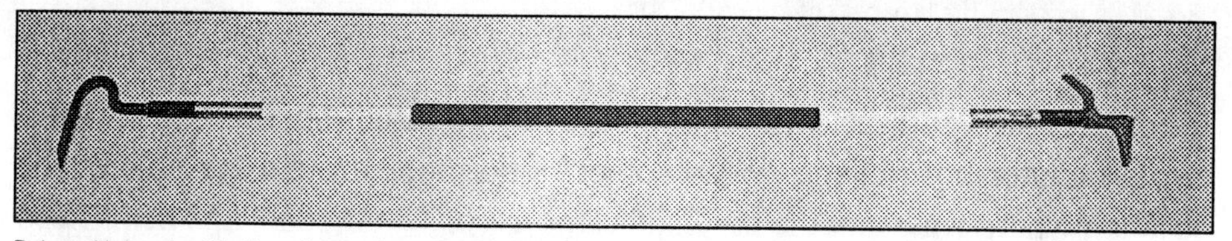
Poles with heads at both ends may look neat, but they are dangerous to carry and use.

Double-headed tools look neat, but for your own safety, stick with a pole with a tool head at one end and a knob, gas shutoff, or other nonlethal appurtenance at the other.

Grips and Miscellaneous Features

Grips—Many poles now come from the manufacturer with some form of foam grip already installed. It usually starts four or five inches up from the bottom of the pole and runs anywhere between 24 and 28 inches up toward the head. These grips are nice but don't last too long under heavy use. They often peel off and really make the tool look ugly. Not that looks are so important—it's just that you lose grip on the tool.

Grips can be added to the pole in the firehouse. Most often these grips will be a little better because you can actually place them in exactly the right position for your hands. French hitching is recommended, as is its simpler counterpart, using friction tape plus rope or wire.

Hook indicator—When using a pole, no matter what type of tool head it has, you always want to be aware of which way the working surface is facing. Once the tool has been pushed through the ceiling, do you know how the hook or pulling surface is oriented? A quick way to solve this problem is to add a hook indicator.

For wooden poles, use a rasp and a file to notch the pole on the side of the hook, down toward the bottom. Make the notches deep enough that you can feel them with gloved hands. Don't make the notches so deep that you ruin the integrity of the pole! Make about four notches, one for each finger. *Do not notch fiberglass poles!*

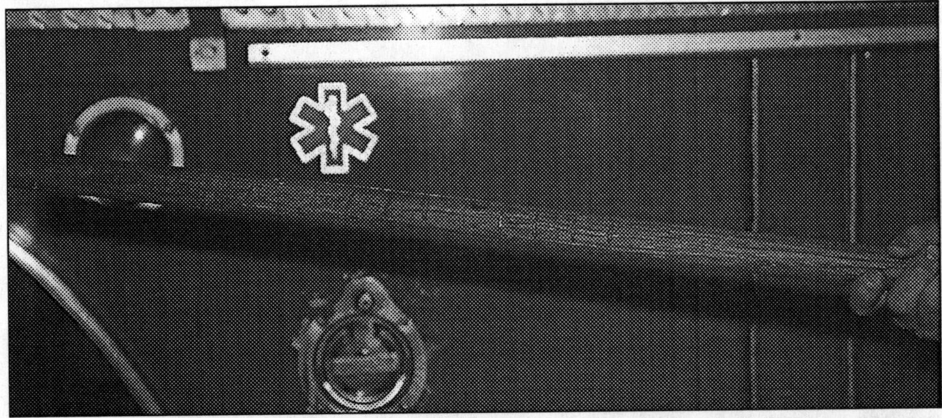

Look closely at the notches in this wooden pike pole handle. They indicate both the location of the hook and the company's signature. This hook belongs to Truck 19, Chicago.

For fiberglass and metal poles, the best way to install hook indicators is to use a radiator hose clamp. Wrap some friction tape around the pole at the location where you want the hook indicator to be. The friction tape helps keep the radiator hose clamp in place. Put the clamp on the pole with the screw assembly on the same side as the hook. Tighten the clamp very tight. Cut off the excess. Wrap the whole thing in electrician's tape to seal it up as best you can. Once installed, you will be able to feel the bump of the hose clamp on the pole and know which way the hook is facing.

Removing and Carrying Poles

Pike poles are dangerous. They can injure firefighters and civilians quickly and severely. There is a specific way that they should be removed from the apparatus and then carried.

Removing from apparatus—Most pike poles are carried in tubes on the fire apparatus, allowing them to be slid in or out with relative ease. It may be the only way that the longer poles can be carried. Some rigs employ a ring and snap bracket, while others will carry a pole upright in a tubular bracket on the running board. No matter how it is carried, a pike must be removed from the apparatus safely.

Poles may be carried in slide tubes on apparatus. Note the selection of hooks available above the ladders of Ladder 10, FDNY.

If your poles are carried in a slide tube, remove them by grasping them by their tops. No matter what type of tool head a pole has, cover the sharp working end with your gloved hand, then slowly and carefully slide it out. Keep a gloved hand over the end of the tool at all times. Be aware of what is around you, and look for anyone in the vicinity. Once the pole clears the tube, immediately invert it so that its working head is pointed toward the ground.

If your poles are carried in a spring-loaded snap bracket and ring, grasp the tool at the top, making sure that your hand covers all the sharp points of the tool head. Lift the tool out of the spring-loaded snap section, then slide the tool out of the ring holder. Always keep the working end of the tool covered with a gloved hand. Once the tool is free of the brackets, invert it and point its working end toward the ground.

If your poles are carried in an upright bracket, grasp the pole about midlength. Raise it straight up into the air until the bottom clears the holder. Set the tool straight down onto the bottom of the pole. Make sure the area is clear, then invert the tool to point the working tool head toward the ground.

Carrying poles—A very natural and comfortable way to carry a pole is to invert it, with the tool head pointing toward the ground and the pole up under your arm, closely tucked into your body.

To carry a pole upstairs, carry the tool head in your hand, your gloved hand covering the sharp points. Keep your arm crooked and the tool head in close to your body, the pole lying out behind you. Drag the pole up the steps, your hand still covering the sharp points. Keep the pole along the wall as you climb the stairs.

Standard Uses

Opening ceilings—There are two basic ways to open a ceiling using any pike pole: the poke method and the throw method.

To execute the poke method, locate the area of the ceiling you are going to start opening. Make sure the area around you is clear, and invert the tool so that the pike head is pointed up. Hold the pole at waist level or slightly higher if you didn't bring a long enough pole. Keep the head of the pole out in front of you with the hook facing away. Don't stand directly underneath the area you are going to open up. With a quick jabbing stroke, drive the tool head completely into the ceiling.

The poke method is very tiring, especially in plaster and lath ceilings.

The throw method for getting the first purchase into a ceiling is sometimes easier on your arms, shoulders, and back, especially in plaster and lath ceilings. Ensure that the area you are working in is clear. Lay the head of the tool out in front of you, with the hook facing up. On tools with two hooks, have the straightest hook facing up. Place one hand underneath the pole in front of you and your other hand on top of the pole, near the handle end. In a quick jerking motion, like a weight lifter, bend slightly at the knees, push down hard with the hand that is on the top side of the

Use the throw method to sink the pike pole head into heavy plaster and lath.

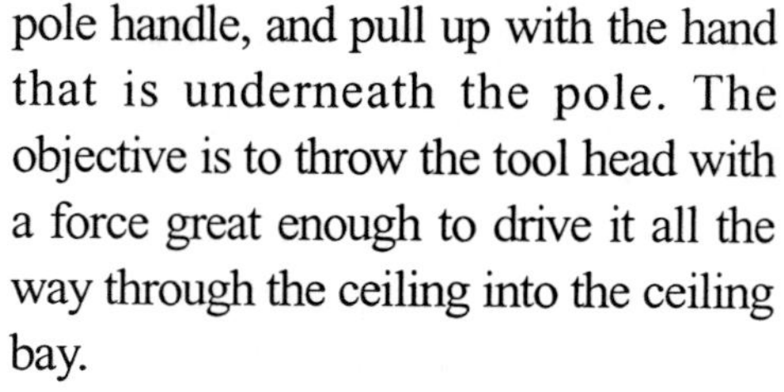

pole handle, and pull up with the hand that is underneath the pole. The objective is to throw the tool head with a force great enough to drive it all the way through the ceiling into the ceiling bay.

Whichever method you choose will work. The throw method isn't really required on ceilings you know to be gypsum board or drop-ceiling panels. Once the tool head is in the ceiling, maintain a solid stance. Turn the hook 90 degrees in either direction and pull down. By turning the hook, you make it grab the ceiling material as you pull; if you don't turn the hook, the head will simply come out of the hole you just made.

Special note: Be very careful when you encounter wire lath ceilings. Falling wire lath ceilings can kill you. It is not uncommon for an entire ceiling area to fall when a firefighter tries to open it. *Do not stand underneath wire lath ceilings.* Use at least an eight-foot hook, and begin working while standing in a doorway or archway. Work until you've gotten a safe area opened up, then move into the room.

Now, look up at the hole you just made. Try to locate the ceiling joist. Once you've located the beam, go to work by reinserting the pike pole into the ceiling, then work the tool along the beam with short, sharp strokes. By following the beam, you shouldn't have to look up to see what is going on. Follow the joist until you get to the wall. Turn around and repeat the process until the area you need is opened.

Opening walls—Wall openings can be started in much the same way as ceilings, either by using the poke method or the throw method.

When using the throw method for opening holes high on walls, the pole does not have to be laid on the floor. Hold it so that the head is even with the wall area you want to drive into. Make sure the area around you is clear. Bring the head of the tool backward, the hook facing the wall you are about to strike. Slam the tool head into the wall with sufficient force to drive it through the material and into the wall bay.

Once the initial purchase has been made, the tool head can be inserted into the hole. The tool head style will then dictate whether the material can be raked out, levered out, or pulled down.

For holes that have to be made low on a wall, hold the tool as if you were spear fishing or throwing a javelin. Jab the tool firmly into the wall to get the initial purchase.

Once the hole is made, insert the pole into the hole, lowering the entire tool head down until the pike either rests on the bottom sill plate of the wall or the pole is at an angle such that the pick is up against the wall material of the adjoining room. Pull back toward you on the pole. The wall material will be removed by prying in short strokes. *Do not force the pole, especially in plaster and wire lath.* Forcing the pole may break it. Remove all of the material.

Opening floors—Don't use a pike pole—use a prying tool. The pike pole is designed to be a push/pull tool. Floors are made of materials that are just too strong to use pike poles to get them open. Axes and prying tools are far superior to the pike pole for this task. Although some of the tool heads we will discuss can be used to pry open floors, always back up the pike pole with a substantial prying tool.

Roof operations—Pike poles are a must during roof operations. Consider taking as a minimum length a 12-foot pole. There will be some roofs that the pole won't fit on, so take an eight-footer. Six-foot poles are of limited use on roofs. They're fine for opening shafts, scuttles, and bulkheads but are too short to push ceilings when a vent hole is cut open. When selecting a pole to take to the roof, know beforehand what job has to be performed. Select a pole that is able to get a firm grab for pulling back the roof material after the cuts are made for the vent hole.

When using a pole to push open ceilings, insert the butt end of the tool, not the working end. More force will be distributed on the ceiling and bigger chunks of the material will fall. By inserting the tool head, you run a greater risk of having the hooks or pulling surfaces snag on rafters, wiring, conduit, or other materials. You won't get the hook out easily, and you may fall into the vent hole if you're pulling too hard on the tool and lose your balance. There are some exceptions to that rule. Some tool heads are less likely to get caught on material and will perform better than if you insert the butt.

Pike Pole Tool Heads

There are many tasks that certain pike pole heads can do better than the other styles. As each type of tool head is discussed, specialty information about how to use it will be added to supplement the standard uses discussed above.

NATIONAL PIKE POLE

Standard Uses

A better name for the national pike pole may be the universal pike pole. This design is truly ancient, dating back to the 14th century, possibly earlier. The head is a standard boat hook shape, with a straight point and a sharply curved hook. The tool head is very flat steel and often has two steel knobs located just above the start of the hook and below the pike point to prevent the head from being driven in too far. Almost every piece of fire apparatus I have ever seen has one or more examples of this type of pike pole.

This style of head is very limited in its capabilities. The tool is designed to actually hook on to the material you are trying to pull. Its very narrow profile makes it the worst tool for pulling gypsum board. The limited dimension of the hook surface penetrates through the gypsum board easily, but rather than grabbing and pulling big chunks, it slices through like a knife. Firefighters must work twice as hard to pull material using this implement. The hook is very sharply curved and makes this tool difficult to use for pulling trim and baseboards. If there isn't much space in the bays with plaster and lath walls or ceilings, the hook will be ineffective.

Don't stop using this pole if this is what you have to work with! By following the steps outlined at the beginning of this chapter, and with practice, you can be very effective with this tool. The national pike pole is a very effective tool for overhaul, ventilation, salvage, and other uses indigenous to the fire service. There are other tool heads that are more effective, but that should not discourage training with and using this tool.

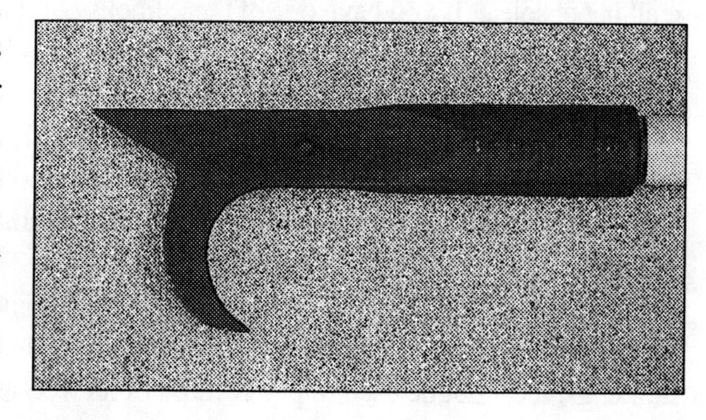

National pike pole.

Learn to be skillful with the tools you have at hand. The national pike pole is a good tool to take to the roof with you. When pushing ceilings, however, insert the butt end of the pole.

Limitations

- The tool head is thin and will not grab large amounts of material.
- The hook is very sharply curved and requires that it be driven in very deeply to grab material.
- It is more difficult to use this tool during overhaul due to the shape of the hook.
- Due to the size and shape of the tool head, it is difficult to use in structures that have limited void spaces behind the walls and ceilings.

PLASTER HOOK

Standard Uses

The plaster hook is another venerable fire service tool. It was the first modification made to a standard pike pole in an attempt to make that tool more efficient for a specific purpose or type of material.

For a good part of this country's building history, plaster and lath was the normal interior finish applied to buildings, both residential and commercial. It proved to be a tough job to remove this material during and after a fire.

The plaster hook is a modified standard pike pole. It has an upright pike, narrow and pointed. Just down from the head, fitted into two slots in the wooden handle, are two collapsible triangular-shaped heads or wings.

When the tool is inserted into the ceiling or wall, the wings retract as they pass through the material. Once inside the bay, the wings snap back out toggle bolt-style, enlarging the pulling area of the tool. The wings are sharpened and bite into the lath on the downstroke, providing good pulling power and removing large amounts of material.

While the tool is very effective, it is outdated. Hook improvements have made it all but obsolete. If you have one of these hooks but no other types, learn to use it. You cannot use the throw method for inserting this hook—its design defeats that. The poke method will allow the tool to function as designed.

You need to notch the handle on this pole to indicate the direction of the wings. I've only seen wooden-handled plaster hooks, so it shouldn't be a problem to make notches. Always know where those wings are to prevent pulling material down on top of yourself.

This tool is for ceilings and walls made of plaster and lath. It is not intended for any other use. It has no prying surfaces to use for removing trim or moldings. It cannot be used to pry because of its wooden handle and its age.

Special Uses

There are no special uses for the plaster hook.

In-House Modifications

If you have anyone in your department who is handy at woodworking, consider replacing the tool's original pole handle with new wood.

Other than that, there are no modifications you can make to this tool. It was designed to pull plaster, which it does, period.

Limitations

There are many tools that are more efficient than this one. Consider retiring it. Put it on the wall above the hux bar.

CHICAGO PIKE POLE

Standard Uses

This tool comes from Chicago and is currently still in wide use by the department there. Originally manufactured in the department shops, the Chicago pike pole is a heavy-duty tool, capable of very heavy fire duty.

The pike head itself is a bit difficult to describe. Here, a picture truly *is* worth a thousand words. The pike is an obelisk shape. An obelisk is a four-sided object topped with a pyramidal shape. This obelisk shape gives the pike tremendous strength and power, and it will penetrate most materials easily.

The hook is also an obelisk shape, at a 90-degree angle to the pike. The hook stays relatively straight along its length, but the pyramid-shaped point angles downward sharply at the end to make an excellent prying and pulling tool.

The tool head attaches to the pole by heavy metal straps that are a forged part of the tool head, rather than a socket. (This is the same way the original pole arms were attached.) The pole is tapered to fit between the straps and then attached to the tool head with screws. This attachment feature makes the tool much stronger when used as a lever for prying, since the stress is more evenly distributed to the pole than by a typical socket attachment. Still, a pike pole is not a strong lever and should be used to lever only light materials during overhaul and other operations.

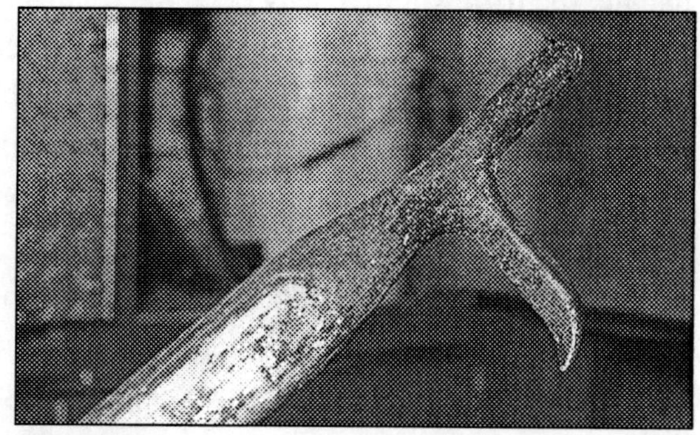

Chicago pike pole.

The 90-degree hook is relatively broad and will grab and pull a lot of material, especially plaster and lath. After this tool head is inserted into a wall or ceiling, turn it at an angle to the material so that the hook will be able to grab as much as possible. The length and width of the hook will distribute your pulling action to a greater area, and more material can be removed in a shorter period of time.

Always be aware of which direction the hook is facing, and keep it angled or turned to pull the most amount of material with each downstroke. Never work with the hook facing you, which will pull material down on top of you. Keep the hook facing left, right, or forward. Also, keep the working end of the tool out and away from you. The size of this tool head makes it easy to follow along a joist.

To remove baseboards, use the pike. If you can, find the area where the last baseboard was installed in the room, which will be the easiest section to remove. Facing the wall, turn the pike pole 90 degrees to your foot. Carefully jam the pike behind the baseboard where the baseboard meets the wall. Start in the corner if you cannot identify the last piece that was installed. Once the pike is well set, pull the pole toward you like a lever. This should remove the baseboard. Sometimes the baseboard is stronger than the wall, and the pike will crush and penetrate into the wall surface, leaving you no fulcrum to pry against. In that event, move the tool along the wall and find a stud. Use the stud to back up the prying action of the tool. If you can't relocate the tool, slide an axe head in between the wall and the pole to give you something to pry against. Remember, the pike on this tool is rather large. Once a little bit of the baseboard has been pried away, reset the tool deeper behind the wood. This is especially true for older-style baseboards that are four inches tall or more.

The angled, pyramid-shaped pulling point on the end of the hook makes fast and easy work of window trim, door trim, and other casings and molding. Baseboards are a little tough with this tool.

To remove window moldings, door trim, and other materials, drive the prying hook down and behind the molding. Pull the pole toward you. The 90-degree hook-to-pike arrangement will make an exceptional fulcrum if needed, but usually the trim will give way once the pole is inserted and pulled.

The use of the tool changes completely when it is turned upside down. You can still use the prying hook to remove finish work, but you will have to be more careful. You will be setting the hook by pulling up and toward you rather than down and away. Don't get overexcited, or you may hook something very near and dear to you if that hook should slip.

With its limited curved hook, the Chicago pike pole should not be taken to the roof by inexperienced firefighters. It works well, but the hook may slip. When using this tool to push down ceilings, insert the butt end of the pole.

Special Uses

The Chicago pike pole is an excellent tool, designed specifically for heavy fire department use. Still, it's just a pike pole. There are no special standout uses for this tool.

Limitations

- The tool head is heavy. It is well balanced but heavier than a standard national pike pole. This becomes a critical factor on long poles, those exceeding 10 feet in length. Long poles may become wobbly and difficult to handle. Firefighters using long poles may lose control of the tool and it will come crashing down. Make sure the area is clear when using long poles!
- The tool head needs to be regularly maintained (sharpened) to adequately penetrate heavy materials such as wire lath.
- The obelisk shape of the pike and hook can inflict serious, penetrating wounds that may require surgery to repair. Take extreme care when carrying and using this tool.
- The hook part of the tool has no real curve. In some circumstances it is difficult to grab material since it slides off the end.

NEW YORK PIKE POLE

Standard Uses

Another important addition to the fire service toolbox also comes from FDNY. The New York pike pole is a standard for all companies in New York City. Designed by the department, this tool is a very strong and massive iron tool that works well pushing in addition to pulling.

Like the Chicago pike pole, the New York pike pole has an obelisk shape, although somewhat more flattened. There are very defined troughs in the iron head on all four sides of the pike, as well as the hook. These troughs actually help to make strong cutting surfaces on all parts of the tool head, which allow the tool to penetrate almost all building materials except metal.

The pike end is a flattened obelisk, with a rather fat but effective chisel point. The hook curves off the main pike point but not severely. The tool head is a socket type, with a continued piece of iron on the rear (side opposite the hook) of the tool head, extending several inches past the socket opening. This extra piece of iron adds strength to the pole for prying. It disperses the force of the tool head more evenly to the pole and helps prevent the pole from snapping at the socket. It will snap, but only if the tool is misused.

The sheer massive size of this tool head really makes it an overhaul-type tool. Although I've described the tool as massive, it's not overly heavy. When

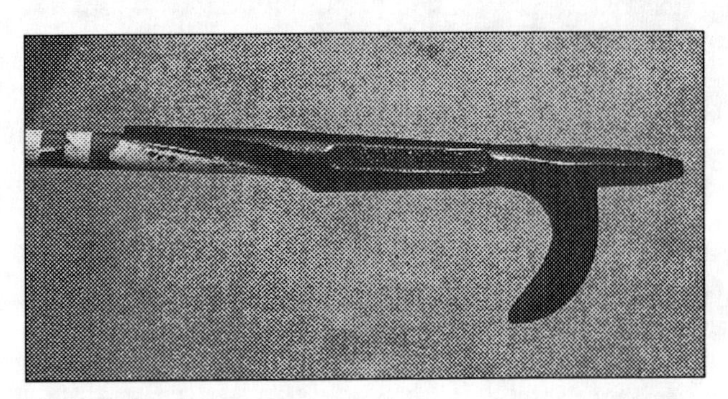
New York pike pole.

compared with a national pike pole, it's heavy, and it's slightly heavier than the Chicago pike pole. The tool is well designed, however, and very well balanced when installed properly. Its thick pike point and fat hook size make it a perfect tool of destruction for overhauling a room, but its use is limited in situations where finesse may be called for to remove baseboards, door trims, and moldings with little damage. The pike will penetrate plaster and lath easily, and the hook is curved enough that it will grab large amounts of material on the downstroke.

Gypsum board and lighter material will also fall easily when using the New York pike pole. The tool head size will grab chunks of plasterboard and make large holes as compared with the national pike pole, which has a tendency to slice through.

The New York pike pole is very effective for pushing materials as well as pulling. The best example of this is FDNY's use of this tool to release drop ladders and counterbalanced stairways on fire escapes. To release a drop ladder, the firefighter should stand under the fire escape if possible. That is the safest location to be in case the drop ladder is defective and falls away from the fire escape. Using the New York pike pole, firmly place the V section of the tool head (created where the hook comes out from the pike) against the bottom of the drop ladder. Maintain pressure and control against the ladder during its entire descent. When releasing counterbalanced stairways, reach up to the metal restraining bar. Insert the tool between the two bars and rotate it until it locks on one of the two bars. Pull the restraining bar clear and let the stairs descend.

The New York pike pole is excellent for ventilating thermal pane windows.

The New York pike pole is an excellent choice for a roof hook. Its curved hook can grab and hold roofing material when opening vent holes. Because it has a hooked head, insert the butt end of the pole to push down ceilings.

Special Uses

It's just a pike pole. This tool is highly recommended for use in older structures that consist of plaster and lath, wire lath, and heavier types of construction.

Limitations

Similiar to the Chicago version, the New York pike has several drawbacks:

- Although well balanced, it is heavier than either a standard national pike pole or a Chicago pike pole.
- The tool head needs to be regularly sharpened to penetrate heavy materials adequately.
- The shape of the pike and hook are dangerous. Exercise caution.
- The massive size of the tool makes it difficult to use in areas that don't require massive overhaul. Baseboards and trim work will usually be obliterated by this tool.

SAN FRANCISCO PIKE POLE

Standard Uses

The San Francisco Fire Department is credited with the development of this pike pole, dating back to the turn of the 20th century. This tool incorporates some of the design features of the Chicago pike pole. Overall, it is an effective and versatile tool.

The tool head is a socket type. The pike is a round shaft that ends in a chisel point. Attached to the round shaft is a triangular hook set at 90 degrees to the shaft of the pike. At the end of the triangular hook is a beveled edge. On the bottom of the hook, a series of serrated teeth give the tool great gripping strength, as well as the ability to grab hold of various materials to make each pull effective.

The chisel point at the end of the pike should be sharpened. Once this is accomplished, the San Francisco pike pole can be used to remove moldings, baseboards, trim work, and floorboards with ease. The chisel point is parallel to the hook, so the hook can be used to set the chisel with your foot.

San Francisco pike pole.

The broad pulling hook and serrated teeth make this tool,

although narrow, very effective for gypsum board. The serrated teeth help hold the material from slipping out from under the flat side of the hook. Use this tool against molding, trim, and baseboard as you would a Chicago pike. The San Francisco hook is also a good choice for a roof hook. Although it has a straight hook like the Chicago pike pole, the San Francisco hook has serrated teeth on the bottom that will help hold the material. The larger hook, however, is a disadvantage when pushing down ceilings. The pike is sharp and will poke holes rather than distribute the force over a large area. Use the butt of the pole to open ceilings from the roof.

The prying end is also an excellent tool for removing floorboards. Insert the chisel in the groove of the floorboard. Step on the hook and drive the chisel in. Push the pike pole handle down, levering the floorboard up. The objective here is to use the pike pole as you would a very large chisel. Make sure that you have a strong purchase in the wood before you begin prying. Also, remember that you are using a pike pole, not a pry bar. Don't overload the tool. The San Francisco tool head is a socket attachment—you don't want to risk snapping it off.

Special Uses

Hey, another pike pole! I know this is getting monotonous, but it *is* another pike pole. It does great things in the hands of a skilled user, but it isn't a do-all tool. One special advantage that this pole does have over the others we've already discussed is that this pole's tool head is very effective for breaking thermal pane glass. The other pike poles will break it also, but the San Francisco pike pole has two very sharp edges, the top chisel and the hook bevel. These enable the tool to penetrate the thermopane glass more easily than broader hooks. This is especially important when breaking glass from an aerial device or ground ladder where tool ability is a bit more important than sheer weight. The San Francisco pike pole can penetrate a pane by jabbing it like a lance rather than by slapping it with its weight. The firefighter using the tool has more control and can use the serrated hook to grab the broken glass and remove a lot of it down to the butyl rubber gasket. The serrated teeth also help to remove the glass shards, making ingress or egress through the window a little bit safer.

Limitations

- The tool head has two sharp edges.
- The versatility of the tool is lost on long poles.
- The tool head needs to be regularly sharpened to penetrate heavy materials adequately.
- This tool head is also an effective weapon. Exercise due caution while using it.

HALLIGAN HOOK PIKE POLE (AKA MULTIPURPOSE HOOK)

Standard Uses

Chief Hugh Halligan designed and developed this workhorse pike pole. Using the premise that a firefighter's tool needs to be as versatile as possible, Chief Halligan combined multiple functions into this pike pole.

It's an odd-looking hook design, resembling a badly misshapen Z. The front hook, and I use that term loosely, is at 90 degrees to the shaft. It is triangular in design, with a flat surface on the bottom and two sloping sides that form a ridge or cutting edge along the top of the tool head. The leading edge of the hook is beveled.

At the point where the front hook joins the shaft, the triangular shapes come to a very sharp point. The point, however, is not in line with the shaft but instead points back at a slight angle. This allows the firefighter to have a penetrating point without having to hold the tool perpendicular to the ground. It also creates one of two fulcrum points for prying.

On the rear of the main tool shaft is another hook. It is designed exactly like the front hook with one exception: This hook angles down and toward the shaft of the tool. It actually looks as if it has been bent; the bend creates another piercing point and fulcrum. This hook, too, is beveled at the end. The tool head is a socket type, but this is a very deep socket. The socket shaft adds strength to the pole, since this tool is designed for prying.

The halligan hook, or multipurpose hook as it is often called, is an extremely versatile tool. Its wide, flat hooks provide excellent pulling surfaces for all types of materials. It should be your tool of choice for opening up tin ceilings. The head is big, so it does require a little extra *umph* to get it into a ceiling or wall, but the triangular cutting surfaces and two penetrating points help greatly.

This tool has both a front and a back hook, so knowing which direction which hook is facing is very important. Depending on how the material you are pulling is constructed, you can use both hooks to pull by keeping the tool head parallel to your body. Keep it angled or turned to pull the most amount of material with each downstroke. "Never work with the hook facing you"

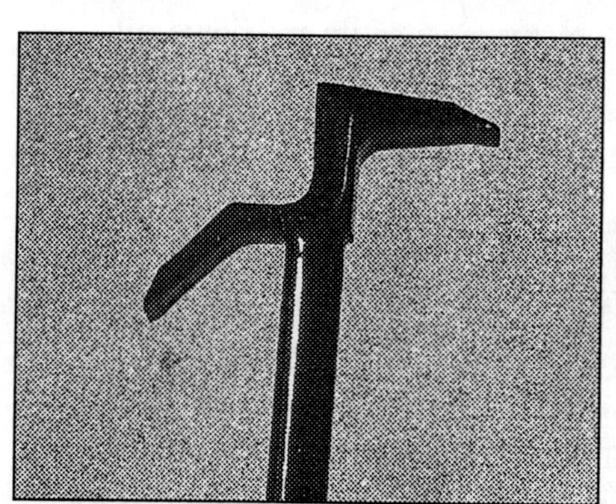

The halligan hook's fulcrum points make it an excellent tool for stripping moldings and trim during overhaul.

doesn't apply here, because there are two. Use the forward, straight hook for pulling material. Try to ensure that the back hook isn't going to pull material down on top of you. Keep the hook turned to the left or right if possible.

The beveled edge on the end of either hook will work against trim and molding. You can easily remove baseboards with this tool. The angled or bent hook will provide an excellent fulcrum but probably won't be needed. Use the bevel on either hook to remove baseboards, depending on the angle you need.

Tin ceilings will open more easily with the halligan hook than with most other types. To open the ceiling, drive the tool straight up into a tin section between two joists. The idea is to bend the metal, not poke through it. Once you bend the metal, the seams should open up. Grab the open seam with the back hook bevel, then lever the ceiling down.

Either hook is excellent for removing floorboards. As described before, insert the beveled end in a groove. In this instance, use the pike pole as you would a pry bar or halligan tool. This tool also has a socket-type head, so be wary of snapping it. Unlike other pike poles, the halligan hook or multipurpose hook is designed to pry, but only light-to-medium materials. Prying metal (other than ductwork or light sheet metal), autos, railroad ties, timber, and other heavy items are all misuses of the tool and may break it. Ordinary construction materials found in residential, commercial, or light-industrial settings are fine. Size up the situation. If heavy prying is called for, get a heavy pry tool.

The halligan hook is valuable on a roof. Its multiple angled hooks, however, will snag on almost anything if you push it through a hole to punch down ceilings, so use the butt end instead. The head of the tool makes a great handle, and it is easier to perform this function by hanging on to the head of the tool.

Special Uses

The halligan hook or multipurpose hook is more than just a standard pulling hook. Its ability to pry and cut makes any such use of it a special use. In a standard configuration—that is, with a fiberglass or wooden handle—this tool is not only a pike pole but also a prying lever. Its strength as a prying tool is limited to light metals, gypsum board, and other such materials. There are other configurations that make the tool much more versatile and powerful.

In-House Modifications

There aren't too many improvements you can make on the halligan hook. You may wish to add a chain link or eyelet to the bottom of the hook, as you would to a halligan bar, for hoisting and ventilation purposes.

Another addition to the tool is to add a prying surface to the butt end of the pole. If you have a tool with a gas shutoff or a ram knob end, a prying tool can be attached. This should be done on fiberglass or steel poles only. Its value when coupled to a wooden pole is very limited because wood will snap under prying pressure.

Limitations

The halligan hook or multipurpose hook is an outstanding tool, but like any other tool, it has its drawbacks.

- The tool is sharp on all sides. Use extreme care when carrying this tool in the firehouse or on the fireground.
- To stay effective, the tool requires a lot of maintenance. Keep its edges sharp, and maintain all the bevels.
- The versatility of the tool is lost when the tool head is on a long pole (longer than 10 feet). Leverage and the ability to pry are diminished.

ROOFMAN'S HOOK

Standard Uses

FDNY firefighters designed and developed this special-use pike pole. They needed a pike pole specifically designed for use by truck company firefighters assigned to the roof position. The specific tasks that roofmen need to perform led to the adaptation of the halligan hook into this configuration.

Like its halligan hook cousin, it's an odd design although extremely durable. It increases the function of the standard-configuration halligan hook. The tool head is attached to a metal pole, which greatly increases its ability to pry. At the opposite end is a curved prying tool that is beveled at the end; it resembles a curved chisel. It isn't a forked end but instead a narrow and curved prying surface. The tool is available only in four-, five-, and six-foot lengths. It is limited in length as a compromise between effectiveness as a prying tool and ease of carrying up ladders, aerials, stairs, and fire escapes.

The roofman's hook can be used as a standard halligan hook or multipurpose hook. Its wide, flat hooks provide excellent pulling surfaces for all types of materials. The head is big and the tool is heavy, so it does require a little extra effort to get it into a ceiling or wall.

The roofman's hook is designed to pry. Doors, windows, bulkhead doors, roof scuttles, shaft hatches, floors, baseboards, hasps, light locks, and all else are no match for it. This tool was designed to be used on a roof. It's a little short for poking down ceilings after the cut is made and the hole opened, however. Consider having a longer hook available for performing that function.

Special Uses

The roofman's hook is really designed for special uses. Its ability to function as a standard pike pole is a real plus. It is a phenomenal tool. When combined with a halligan bar or similar prying tool, the roofman's hook can be used as a driving tool.

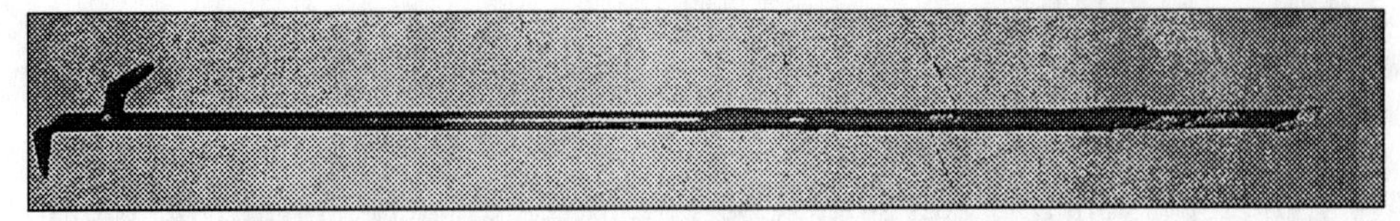

All-metal roofman's hook.

For inward-swinging doors, one firefighter can make quick work of even a well-secured door. Drive the hook of the halligan or similar prying tool completely into the doorjamb six inches above or below the lock. Use the roofman's hook as the driving tool. To do this, invert the roofman's hook so that the head is on the ground, and turn the tool so that the flat, straight hook is facing you, with the bent hook to the rear. Place your foot on the flat side of the hook, and hold the roofman's hook so that it is up on the fulcrum point at the very top of the head. While holding the halligan or similar tool with one hand, hammer the hook of the tool into the door frame with the shaft of the roofman's hook. You should strike the bar firmly. Adjust it as required to continue striking until the hook of the pry bar is driven in sufficiently. Push the pry bar down and the door will open. Otherwise, the adze may be driven between the door and jamb like a standard halligan.

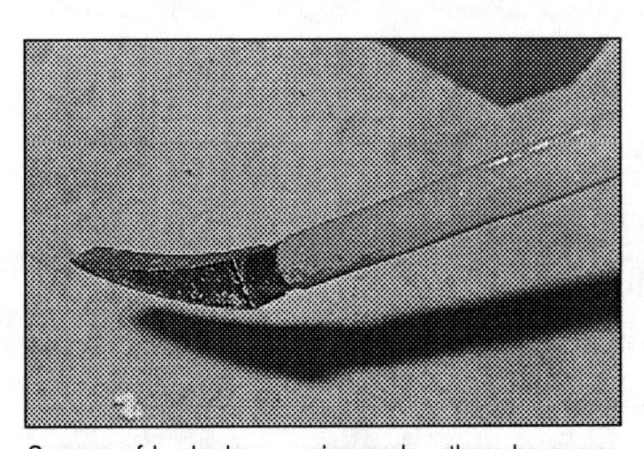

Some roof hooks have prying ends; others have ram knobs. Prying ends add versatility.

The fact that the roofman's hook is an all-metal tool truly enhances its capabilities as a prying hook. This tool really fits the bill as having true multiple uses.

Limitations

The roofman's hook, like any other tool, has its drawbacks.

- The tool head is sharp on all sides. Use extreme care when carrying it up ladders or fire escapes.
- The tool is sharp at both ends.
- It's heavier than a standard pike pole. Use extreme care not to let this tool (or any other tool) fall from the roof.
- Lots of maintenance is required on the edges and bevels.
- It is not available in lengths greater than six feet.

MULTIFUNCTION HOOK

Standard Uses

The best description for the way this tool looks is bizarre. The multifunction hook is an adaptation of the standard halligan hook. Its capabilities are somewhere between the standard hook and the pry bar.

The tool head resembles a badly misshapen Z. It's so badly misshapen that it looks like a steamroller ran over it. The front hook is at a 90-degree angle to the shaft; it is flat and is really an adze tool. The leading edge of this adze is beveled slightly, and the top side of the adze slopes back toward the tool shaft and ends at the top of the tool head in a broad point, which is used as a strong fulcrum for prying. The forward broad adze hook is more angled than a standard halligan hook head, which increases its leverage capabilities. There is a small striking surface area located on the back side of the forward hook, between the top of the tool and where the back hook is attached. This area can be used to drive the adze hook into an area for a greater purchase and greater pulling power. Remember, this is still a pike pole! Do not get the tool head situated into such an area that you cannot get it removed. If it gets wedged too tight, you may snap the handle trying to pry it out. The tool is designed for light to moderate prying needs.

On the back side of the main tool shaft is another hook. This hook angles down and toward the shaft of the tool. It actually looks as if it has been bent. The bend creates another piercing point and fulcrum. This hook is also beveled at the end. Like the front hook, it is relatively wide. It retains more of the triangular shape of a standard halligan hook.

The tool head is attached to the pole by a socket arrangement, which decreases the tool's ability to pry. Prying strength will be more dependent on the pole material selected than on the capability of the tool head.

The multifunction can be used as a standard halligan hook or multipurpose hook or used instead of a roofman's hook. Its wide, flat hooks provide excellent pulling surfaces for all types of materials. The head is big and the tool is heavier, so it does require a little extra effort to get it into a ceiling or wall. The capacity for prying with this tool head is tremendous.

The hooks on this tool are very broad and will pull down much larger chunks of material with each stroke. Really concentrate on not standing directly underneath the section of ceiling or wall you are working on when pulling with this tool, and don't let anyone else stand there, either. This tool does have more prying capabilities than a standard configuration pike pole but still less than a pry bar.

The broad, adze-type front hook on the multifunction hook is outstanding for removing baseboards. The additional leverage added by the pole makes the multifunction hook easier to use for trim removal than a halligan bar or similar tool. The broad, flat surfaces of the tool allow the firefighter to have a firmer hold on paneling, plywood sheeting, and other large materials. It is great for tearing away boards from windows and doors on vacant structures. Either hook will work against floorboards.

The multifunction hook is designed to pry. It will pry a variety of materials and will work well as a prying tool in a multitude of situations. I cannot stress enough, however, that the tool's strength will be limited by its weakest part, which will be the pole material itself. Ordinary construction materials found in residential, commercial, or light industrial settings are no match for the multifunction hook. Size up the situation. If very heavy prying is called for, get a heavy tool. If there is a life-safety situation, get a prying tool!

Special Uses

The multifunction hook is designed for special uses. Its ability to function as a standard pike pole is a real plus. It is a good tool. It can be used to a limited extent for forcible entry other than breaking windows. The large adze-type front hook may be able to pop doors open. This will depend more on the ability of the firefighter to maneuver the tool into place because the pole will be in the way. It will work but not as efficiently as a pry bar.

The real special advantage the multifunction hook has is as a roofman's tool. The broad prying surfaces make this tool very useful for opening scuttles and skylights, shaftways, elevator doors, and bulkhead doors, for example.

This tool combines the usefulness of a standard halligan, a halligan adze, and a roofman's hook all into one. It is a great addition for suburban and rural fire departments where forcible entry and roof work are not quite to the extremes that they are in urban areas. This tool will make it very easy to open many of the skylights and windows found on many modern single-family residences. It also makes short work of the vented ridge caps on roofs of single-family dwellings. The large head of the tool should not be inserted into vent holes to knock down ceilings because there is too great a risk of its getting snagged. Use the butt end of the pole instead.

Limitations

The multifunction hook does have several limitations.

- The tool head is sharp on all sides.
- The tool is sharp at both ends.
- To stay effective, it requires a lot of maintenance. Keep its edges sharp and maintain the bevels.
- Because of its ability to do so many things, this tool is prone to being misused on the fireground.

DRYWALL HOOK

Standard Uses

The drywall hook is another tool designed by an FDNY firefighter. This style of tool is fast becoming a favorite of firefighters because they can do massive amounts of work with it with little effort.

The tool is well designed. It's a funny-looking thing, but it immediately becomes obvious that, when used properly, it will put most other pike poles to shame. It has a flanged head that is angled up from the shaft. It then bends back down, and the four-inch-wide pulling surface ends in a series of teeth. On the top of the tool flange is a beveled fin with a cutting surface. The fin is angled so that it has an offset point for piercing into materials. On the bottom of the fin is a small, sharp hook for getting a purchase in tight areas or to use as a fulcrum for the cutting fin. The tool will make short work of drywall. This is a true pulling tool.

To use the drywall hook to open a ceiling, get an initial purchase with the tool. The massive size of the head will send it sailing through gypsum board if you use the throwing method. The sharp cutter on the top of the tool will easily pierce gypsum board if you use the poke method. Either way, the tool can be put up into the ceiling easily.

Always work with the large, serrated tooth side of the tool away from you. Any force you apply will be distributed over a very large area of the gypsum board and will open it up in big sections. Don't stand underneath the area you are working on. Apply enough force to the handle of the pole to try to pull the plasterboard over the nails.

Follow the joist to pull the material down. Use the teeth to grab and rake insulation out of the bays. Stay conscious of the seams of the gypsum board, which will sometimes break at the seams and come down in entire sheets. I've

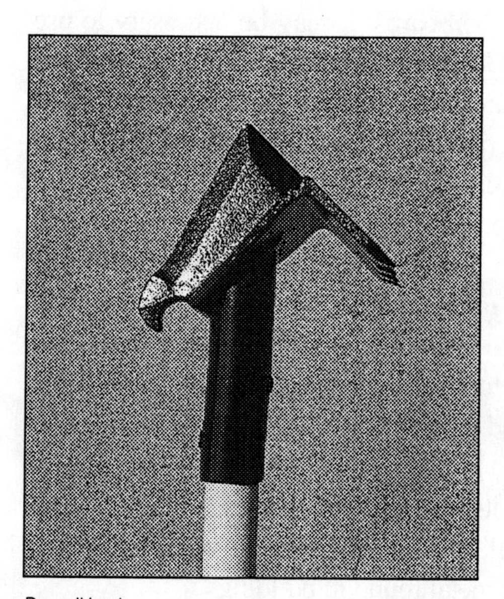

Drywall hook.

also seen it hinge at the tape seams and fall off the ceiling like an opening door and smack guys in the face. When using the drywall hook, don't go crazy. The tool will do the work with little effort on your part. Stay alert to the area, and recognize how you are taking the material down so that it doesn't fall on you.

The drywall hook will also devastate plaster and lath. The broad pulling surface and serrated teeth grab hold of several laths at once and will really make short work of such ceilings. The sharp penetrating point of the top fin also helps penetrate the difficult wire lath material. Use the tool as you would any other pike pole—the difference will be in the amount of material you pull down.

The drywall hook is invaluable in opening up walls. The wide pulling surface will quickly rake out large pieces of the wall. You can set the tool two ways—the poke method or the throw method. Either way will get you into the wall so you can go to work. Use the tool more like a rake than a pulling pole. Get yourself into a comfortable position, then pull the tool down and out, raking the material away as you move along.

This tool has a very wide pulling surface, so be careful if it hangs up. If you feel resistance on the pull, check to make sure you're not hooked over conduit, water pipes, wire, or anything else that may be hidden in there.

The serrated edge on the end of the angled pulling surface, combined with its width, works well on trim and molding. Baseboards are somewhat difficult to remove with this tool as compared with others. The fin on the top will provide a fulcrum, if needed.

Leverage can be adapted to the situation when using the drywall hook. It will pry, although it wasn't primarily designed to be a prying device. This tool does have more prying capabilities than a standard-configuration pike pole but less than a halligan hook or similar tool. For heavy prying, the appropriate tool is a pry bar of some kind.

Use the serrated edge for removing floorboards. It may be necessary to use another tool such as an axe or bar to make the initial purchase into the floor so that the serrated tooth edge can get a purchase on the edge of the floorboard or sheeting material. Whether you push or pull the pole will depend on where you are standing and which way the floor must be removed.

Rake the tool over the surface until you have a good purchase on the edge of the board. Don't stand on the material you're trying to remove. Step to one side, and push the pole forward until it rocks up onto the fin on top of the tool. This will create the fulcrum you need to lever the board out of place or to pull the sheeting up. In most cases, another tool will be needed to complete the job, since pulling floors with a tool designed to rake drywall and plaster and lath is not very effective. Get a pry bar of some kind.

The name of this tool hides many of its actual effective uses. Although other types of tool heads work well in metal, this one is especially effective when working with ductwork and other sheet metal found in buildings.

The fin-shaped cutting tool mounted on the top of the tool will pierce sheet metal ductwork with ease. Once the fin opens a hole big enough to allow the tool head to be inserted, the broad tool head will distribute your pulling force over a large area and allow you to pull down the metal.

The drywall hook is also very effective at pulling tin ceilings. Strike the tin ceiling hard with the top of the tool to cave it in a little. This will open up the seams of the panel you are trying to pull. Insert the serrated teeth into the seam and pull down. An additional feature that makes working the head in metal effective is the small hook that is attached where the back of the fin joins the main tool head. This little hook can be used to make a can opener out of the tool.

For example, let's say you need to open some commercial-grade ductwork in a restaurant. Use the top point on the fin to poke a purchase hole in the duct. Insert the small hook and pull backward to set it. Now, push forward on the pole and pivot the fin up into the ductwork. Reset the tool about midway down the new opening at about 90 degrees from where you originally started. Set the hook and repeat the process. It will slice open the ductwork, allowing you to push the tool up and through the sliced area to get a purchase with the serrated edge.

The hook can also be used to pull materials such as trim in areas where the full-sized face of the tool won't fit. The entire tool head tapers to the back toward the hook and will allow you to grab materials and get it started.

The drywall hook is a valuable tool for almost all types of construction. It is a good choice for roof operations. It should not be used to punch down ceilings, however. Use the butt end of the pole instead.

Special Uses

There are limited special uses for the drywall hook, but the one that stands out the most is its value as a tool for creating delicate cuts in drywall to create inspection holes that can be easily patched.

If an inspection hole must be created in an area that has had no fire involvement, a firefighter can manipulate the hook carefully and precisely to create a small hole. Use the fin on the top side of the tool. Grasp the tool up close to the head. Push the fin into the drywall and pull down. The fin will slice the drywall neatly. Make three other cuts using the same technique. Carefully use the serrated edge to pry out the cut square of drywall. An inspection for fire can be made, and the owner can easily patch the hole.

The broad head of this tool makes it an effective rake for overhauling debris and stirring up trash in dumpsters for extinguishment. The tool is very efficient at removing window glass and frames during ventilation.

Limitations

- The top fin of the tool is very sharp. Observe safety practices.

- The tool is heavy due to the massive size of the head.
- It doesn't lend itself very well to prying heavy materials like flooring.

EK (E-KAY) HOOK (AKA EKERT HOOK)

Standard Uses

Because the subject of pike poles that can cut metal came up in the drywall hook section, the EK hook is a logical tool to follow.

The EK hook is designed to cut open metal ductwork, tin ceilings, and other sheet metal. The tool head is a flat metal cutting head that extends straight up from the shaft socket. The head curves out from the shaft to form a crescent-shaped blade surface on the top. The bottom side of the crescent is actually the hook part of the tool. Notched into the bottom side are sharp, serrated teeth.

The main use of the EK hook is as a cutter for opening ductwork in restaurants, fast-food places, and commercial and residential buildings. The tool can also be used to cut open tin ceilings, metal pole buildings, and other places where light metals are found. The head is curved and forms a hook on the bottom side so that it can be used as a conventional pike pole, but its use is limited to specific materials. The sharp surfaces that make this an excellent metal cutter work against the tool when it is used to pull gypsum board. The thin head design and sharp surfaces slice through gypsum board on both the upstroke and the pull stroke. Even when the head is turned, the tool slices its way out without really grabbing much material. It's a downside to the tool. Remember, though, that this tool was primarily designed to cut, not pull. Its head does work effectively in plaster and lath, especially in wire lath.

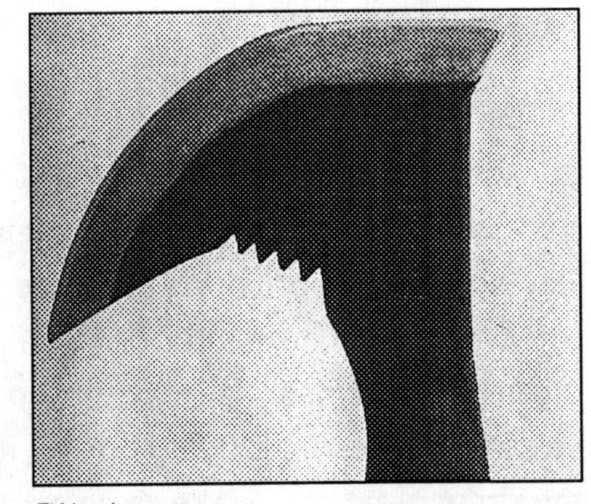
EK hook.

The tool is basically used like a long can opener. The sharp tool head will pierce the metal when you insert it, and the teeth will cut again when the tool is pulled back. The initial opening can be made using either the poke method or the throw method. The throw method is a bit more dangerous because of the sharp head, and the tool has a tendency to bounce back if it doesn't get a purchase on the first throw.

Use the flatter part of the cutting head, the section more in line with the shaft, for the poke method. For the throw method, use the sharp points of the tool. There is a very sharp point on top of the head. There is another, even sharper cutting/piercing point at the apex of the tool hook where the top cutting blade meets the bottom cutting surface. To use this point, throw the tool with the point up toward the duct or metal. How it is used after the initial opening is made depends on what you are opening and how far you have to open it.

Ductwork can be found in every structure having HVAC systems. It presents a formidable obstacle to pike poles like the national pike poke, New York pike, San Francisco hook, and others that are designed to pull standard building materials.

For ductwork, the tool blade can be placed in the cut of the initial opening, then pushed along like a knife to open the metal. You may have to jab the tool to make the cut. These methods work for light metal. When you run into doubled metal such as a seam or joint, the tool may come to an abrupt stop. Try jabbing harder, or cut around the seam or joint if possible. Only open a hole the size you need to do the initial job you set out to do. If larger amounts of duct have to be opened, use tin snips or, better yet, an air chisel. The EK hook's teeth can also be used for cutting by pulling back on the pole and opening up the metal.

When opening metal ductwork, try to follow along a seam support for the metal. Don't cut the seam; try to follow a path parallel to it. This will provide support for the cutting action of the tool. The metal will flex and bind easily, so following along a seam is like following along a joist when pulling conventional materials—it offers more support and acts as a guide.

You may be surprised by the number of tin ceilings you can encounter. In urban areas, old neighborhoods are being revitalized and restored to their original appearance. In suburban and rural areas, tin ceilings that have been scavenged from urban buildings get new life in new private homes or specialty shops. Having an EK hook available to you may be a great advantage.

Tin ceilings are usually very difficult to open. Falling tin sections have razor-sharp edges. Having the EK hook and other tools available will save you from some terrible cuts and lacerations if the material can be handled completely with tools. The sharp cutting blade is an advantage when opening tin ceilings. Using the tool, the ceilings can be pulled down or an opening made to get a good purchase with standard pike poles. The sharp point on the end of the curved hook is great for searching for and opening up seams in tin ceilings. Also, look for purchase points in areas where pipes pierce the ceiling or where openings have been cut for ceiling fixtures.

Metal buildings, most commonly identified with rural areas, are now common in suburban and urban areas. Storage buildings, metal shops, body shops, and small convenience stores are commonly metal buildings. Additionally, rural areas are seeing ever-increasing numbers of this type of structure for machine sheds, shops, and even barns.

The EK hook can really be advantageous for getting into the sides of metal buildings. It will also pierce and cut mobile homes and aluminum siding. Using the tool for this purpose is similar to opening ductwork. Get an initial purchase hole either by poking or throwing the tool head through the metal.

Remove the tool and reset it with the blade following the split. Push the tool along, cutting as you go. You can also cut by pulling down on the tool, letting the teeth do their job. I think the best description of what you are doing is cutting the metal the way you would cut paper with a knife. It isn't as easy as that, but you get the idea. To follow along attachment points of the metal siding, look for the rivets or screws, and follow along as best you can. Don't cut on the seam; rather, cut adjacent to it. The curve of the tool provides enough of a hook to snatch and hook insulation easily and to pull it out of the opening. Open a hole big enough to accomplish your objective. This is not the tool to use to open a building for a rescue. When speed is required, get a power tool. It is difficult to use this tool during overhaul due to the shape of the hook and the bottom-side cutting teeth. It has a very narrow cutting head and an extremely sharp curved head. It will not grab large amounts of material when pulling, and it cuts on the way in and cuts on the way out, making small, narrow slits. The tool requires greater effort to remove basic construction material during overhaul. Use a tool that has a greater pulling surface.

The EK hook may be of value as a roof tool or a tool to have with you in a bucket or on an aerial. The EK hook will help you remove the tin from tinned-up windows, ductwork, shaft covers, or other light metals you'll find on a roof.

This is a valuable tool to have in your inventory. It is a limited-use item, however, because it was designed to open metal. It will work in some instances as a standard pike pole, but it is really lousy when used on gypsum board. Don't bother taking this tool to the roof except in the special situations described above. In most cases, this tool is ineffective.

Special Uses

There are no true special uses for the EK hook. It is designed for those situations involving sheet metal that cannot be handled by standard pike poles.

Limitations

- The cutting edges of this tool are sharp and will inflict severe injuries if not properly handled.
- It's a limited-use tool. It performs poorly when used on drywall.
- When piercing ceilings or walls, this tool will cut wires and other electrical components, creating a shock hazard.
- It will damage copper water pipes if you hit them with the cutting blade.
- The tool head is thin and will not grab large amounts of material.
- The hook is very sharply curved, and the head must be driven in very deeply to grab material.
- It is difficult to use this tool during overhaul due to the shape of the hook and the bottom-side cutting teeth.

- The size and shape of the tool head make it difficult to use in limited void spaces within walls and ceilings.

BOSTON RAKE

Standard Uses

The Boston rake isn't really a pike pole—it doesn't have a pike or a hook. The Boston rake is a raking tool designed to pull plaster and lath. Although it will function as a pike pole, albeit poorly, the Boston rake is a superior tool for use in older-construction buildings that have heavy plaster and lath, wire lath, heavy baseboards and trim, and large window trims and frames.

The rake is a gooseneck-shafted tool that flattens out on the curve and develops into a flat blade that widens toward the end. The leading edge of the flat tool is angled back at about a 45-degree angle. It is beveled, and the bottom of the angle is very sharp. The head is attached to the pole by a socket.

Boston firefighters are heavy users of this tool, and for them it is very effective. Boston is a very old city, with construction that dates back to the 17th century. The rake works well in older areas of the city where plaster and lath is the predominant building material.

The rake does not work well when you use the poke method. There is no pike, so the top side of the tool has no method for piercing into any material. You can turn the tool around and use the ram knob or the gas shutoff knob to make an initial hole in the ceiling or wall. The rake works better by using the throw method and inserting the tool by using the leading angled edge.

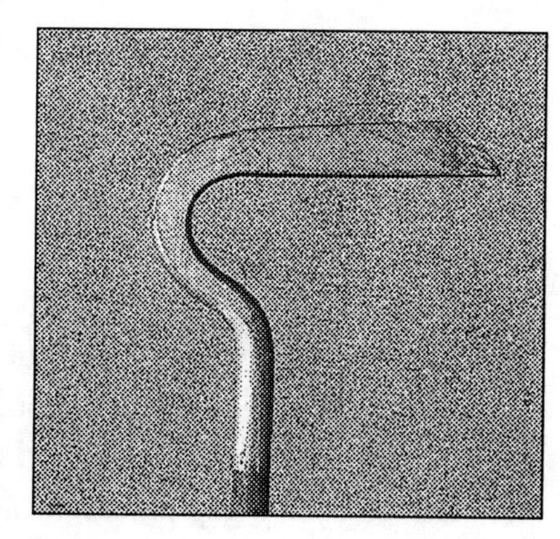

Boston rake.

Once the hook pierces the ceiling or wall, push the tool forward or down and into the void space. Ensure that the head is facing away from you, then pull down on the pole for ceilings or straight out when pulling a wall. The length of the tool's flat blade will distribute the force being applied to a large surface area, and much material will come down.

Once you've opened up an area and can see, continue to work the tool in the same manner, but stay very close to the joists or stud. The rake is excellent for pulling out the lath, nails and all.

The rake doesn't work really well in drywall. It has the same problem that the national and other thin-headed poles have: It has a tendency to slip back or slice through the gypsum board without really getting a firm hold.

This can be overcome by being more conscious of where you're using the

The Boston rake efficiently removes windows for ventilation.

tool. When trying to remove gypsum board, whether on a ceiling or wall, try to follow the seams. Work the tool along the seams and pull the gypsum board past the screws or nails. You'll be more effective with that method.

The wide, flat head of the tool is effective for removing trim, but it is a bit clumsy. To remove baseboard, let the head of the tool slide along the wall, and drive it into the gap where the baseboard meets the wall. Let it slide with enough force to set the point of the beveled edge of the tool quite deep. Start in the corner if you can. Pull the pole toward you. The widest part of the blade is at the tip just behind the angled edge. By pulling the pole, you are forcing the blade to open a gap at least as large as the width of the blade. This may be sufficient to pull the nails.

Window and door trim can be removed in much the same way. Slide the flat of the rake behind the trim and lever the pole toward you. As the tool turns, the blade will force the trim off the wall. When used carefully, the Boston rake will remove baseboards and trim with little or no damage. That's an important PR fact to remember when searching unaffected rooms.

Special Uses

The Boston rake is a special-use tool in and of itself. There are some applications, however, at which the Boston rake can outperform some other standard hooks.

The primary special use is for breaking thermal pane windows. The length of the tool head and its beveled leading edge and point make breaking the newer thermal panes a cinch. Broader pike pole points have a tendency to bounce off the glass, especially the newest types of windows that have the inert gas between the panes for energy efficiency. The rake head can be used to pierce the glass rather than just smash it. The length of the blade will penetrate both layers with relative ease. There may be less flying glass because you are using a narrower tool to assault it.

Once the glass is broken, the gooseneck shape of the tool head makes cleaning out the remaining shards and removing the entire frame easy. Removing the frames of the thermal pane windows is your best option. The butyl cement that is used to hold the glass in place makes it impossible to clean all the pieces out of the frame.

The Boston rake can also be used on tin ceilings and other light metals. The point of the tool can be driven in, then the metal levered from its position. The rake lacks a good grabbing hook, so it should be used in conjunction with another type of pole that can grab and pull the metal. The Boston rake is good for opening up purchase points for other tools and is a good choice for roof operations.

When using the Boston rake on the roof, make sure you have a 12-foot handle. The deep curve of the tool head will hold roofing materials and pull them back. This tool can be inserted into a hole headfirst, and the large surface of the blade will distribute the force over a larger area. It may snag, but if you know which way the head is facing, you may be able to get material to slide off the end of the hook, out and away from the gooseneck. Try using the head during practice sessions. Don't experiment during a real fire when lives are at stake.

Limitations

- The Boston rake has no pike.
- It is unable to actually grab on to material.
- Its thin head makes it difficult to use in gypsum board.
- It can easily get snagged on wiring and conduit.

CLEMENS HOOK™

Standard Uses

The clemens hook is a modernization of the traditional idea of a pike pole. It is designed primarily as a push/pull tool for use during overhaul operations.

The tool head is a socket-type head that resembles a misshapen crescent moon, pointed at the top. It widens out to form an adze-type tool at the bottom of the crescent. The adze has a bevel. Running up the center of the crescent shape is a thin fin used to cut through material. Located on the shaft just below the tool head are two lugs to prevent the tool from being inserted too far into the material to be pulled. They don't work, but they're there. The pointed end of the tool can be sharpened to be an effective pike for standard overhaul operations. The beveled end on the adze can also be sharpened and is effective in opening up walls and trim. The clemens tool doesn't have a hook per se as do standard boat hook-style pike poles, but the angle created between the bottom adze end of the tool and the shaft creates a notch that can be used like a hook. The lack of a hook makes this tool less likely to snag on wires, conduit, plumbing, and other surprises hidden behind walls and ceilings. The tool is much more efficient at making openings on the push stroke than it is on the pull stroke. The poke method is the best way to set this hook into the material because the fin and widening head will smash and open the material as you push it in. This is very effective on gypsum board, but you'll break a real sweat in plaster and lath. The tool head will smash and break the laths and plaster, but you'll have to supply the extra effort.

An interesting way to use this tool is to turn it before you pull. Drive the tool into the wall or ceiling with the head parallel to your body. Once it is into the void area behind the wall or ceiling, turn the tool 90 degrees. Pull. The tool should make a bigger hole on the downstroke. Not always, but a good percentage of the time this will work. Repeat the process. Without a hook, the clemens tool is poor at pulling ceilings. The hook section located at the angle created by the adze end and shaft will always be turned toward you when you pull ceilings. Beware.

The angle of the beveled adze is very good for removing trim. The shape of the head will give you a good fulcrum when using the adze side of the tool. The fulcrum is poor for the point side. Turn the face of the tool against the wall and rake it downward, driving the beveled end of the adze in and behind the trim. Lift up on the pole, and the trim should pop off.

Use the pointed end of the crescent to remove baseboards. Carefully jam the point end in and behind the baseboard where the baseboard meets the wall. With this tool, the fulcrum is poor for the point side, whether the wall gives way or not. If this is the case, move along the wall until you find a stud. Use the stud to back the tool, and the baseboard will be removed easily.

The clemens hook isn't a replacement for a standard pike pole, but it does have its place. It is excellent in areas that are primarily gypsum board and have standard wood or plastic trim. It shouldn't be your first choice as a roof tool, since it was designed to be a push/pull tool. It has very limited grabbing capacity, and the roofing material may be too thick for it to grab.

It does have an advantage during roof venting, though. The tool head has very small "hooks" or angles that can be used to snag things, while the crescent shape of the head will allow material to slide by it without becoming too terribly snagged. The broad face and cutting fin will distribute force over a wide area and push down a lot of ceiling. It can be used on the roof but should be limited to use by experienced roofmen.

Limitations

- It has no real hook to grab material.
- It is not conducive to prying.
- Its massive head is difficult to get through a lot of building materials without your really having to work at it.

L.A. TRASH HOOK (AKA ARSON-TRASH HOOK)

Standard Uses

Here is a tool whose name only hints at its serviceability on the fireground. Used heavily in the Los Angeles area, as well as in numerous departments in other towns and cities, the trash hook has become a very versatile tool, much more useful than just for raking burning trash. Some tool manufacturers have had trouble with the name, so may call their product an arson-trash hook. Don't let the name of the tool fool you—this is a great tool.

The tool head is an upside-down triangle. The apex of the triangle is attached to the socket. At the two opposing corners of the triangle at the top are attached two six-inch-or-longer metal prongs. These prongs are cylindrical and have conical points.

The standard use of this tool is to rake trash. The two prongs penetrate into the muck and allow firefighters to turn over smoldering debris to be washed down and thoroughly extinguished.

A minimum of a six-foot pole allows firefighters to stay out of the debris while working. This is extremely important because of the unknown materials that can be found in trash. You use the tool for this technique as you would a lawn rake at home. Grab the material, pull it, and turn it over. Simple. It works great. It allows more material to be turned over with each stroke of the tool. Narrower hooks cannot move material this easily, so you spend more time standing in the stinking mess trying to put it out.

That's it. That's the standard use for a trash hook.

Special Uses

Firefighters discovered that the trash hook was an excellent multiuse tool in the field. Through trial and experimentation, the tool has become a valuable asset for ventilation, overhaul, arson investigation, and roof operations.

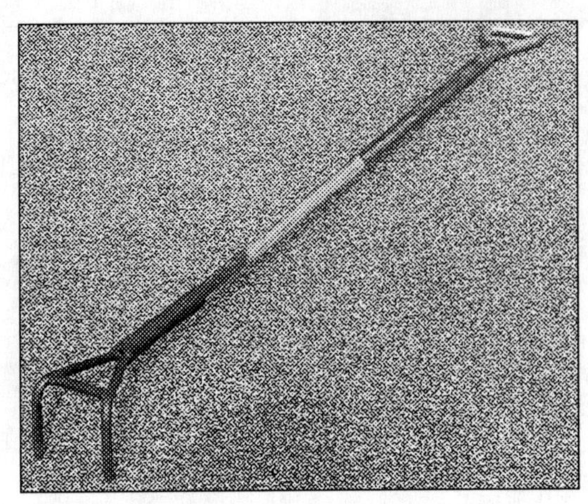
L.A. trash hook.

The sharp tines of the rake will penetrate and shatter glass without any problem. By using the throw method, a firefighter can shatter windows, including thermal panes, with ease. Once the glass is broken, the wide head of the rake will make short work of removing all of the remaining window sash. This gives firefighters an area clear of glass shards and a large, clear area for smoke and fire to vent.

During overhaul operations, a firefighter using a trash hook can open large areas with the least effort of any pole tool. The awkward location of the tines, however, makes gaining the initial purchase a little more difficult. To open ceilings with the trash hook, use the throw method.

Once the trash hook head is into the bay, turn the tool so that the tines are facing away from you, then pull down and away. Know what material you're in, though, because you'll have to gauge how hard to pull. If you pull too hard on gypsum board, the tines may pull right through. Be on the lookout for plasterboard that may pivot on the seam and slap you in the face. Once your initial hole is opened, continue raking the area clean, following the joist. The very large surface area of the trash hook head will pull down large amounts of debris. When you get to the wall, turn around and repeat the process. Because of the amount of material you will be pulling down at once, make sure you don't leave any tools or equipment lying on the floor. You will soon bury them in debris. The trash hook makes quick work of plaster and lath, as well as gypsum board. It is almost useless in wire lath, and the power of the pulling surface may cause you to pull the whole ceiling down on you at once.

Opening walls with the trash hook is an operation similar to opening ceilings. In walls, though, you can use either the poke method or the throw method. The throw method is easier. Once the tines have pierced the wall, lift up on the pole so that the tines are parallel with you (vertical) inside the wall. Pull. For gypsum board, pull gently; for plaster and lath, pull hard. You will rake out a lot of material. Take a look at where you are, and reset the tool along the stud. Go to work. You won't be there long. Be conscious of any resistance you feel while pulling. The long tines of the hook may snag on wires, conduit, or plumbing. The wall material should offer no resistance to the tool.

The trash hook is not a do-all type of tool, however. Its head design makes it awkward for certain overhaul functions, such as trim removal and baseboard removal. It works, and you can perform the job with the trash hook, but it's a

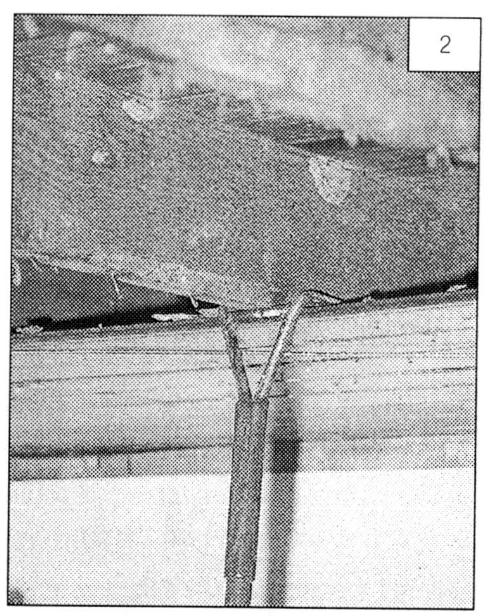

1. The trash hook will easily remove wooden flooring and ceilings. First, remove one board with conventional tools. 2. Next, insert the tines on either side of a joist. 3. Pull up on the tool handle. 4. The tines will force the board from the joist. 5. Once it is loose, pull down hard to remove the entire board.

little bit more time-consuming and can sometimes be frustrating. Use a different hook for these jobs, or have a halligan-type pry bar handy.

The trash hook is an advantageous tool for the arson investigator. Unlike the shovel, the trash hook can pick through debris while evidence is being sought. The trash hook, when properly used, will move material without jumbling it up the way a shovel does.

The trash hook is a great tool to take to the roof with you. The large tines will grab and hold roof material from the vent hole and help to pull it back out of the way. The triangular head will slip by most obstacles inside the hole, and the very large head and tines will push down tremendous amounts of ceiling material. The tines may snag when retrieving the tool from the hole, but if you are conscious of which way the head is facing, you can angle it and allow the material to slip off. You must know which way the tool is facing to be effective; adding a hook indicator to the handle will help.

The L.A. trash hook is a great tool to have. It is a true multipurpose tool and very effective in many different situations. The fire service needs to come up with a better name, though.

Limitations

- Because of its name, the value of the tool is overlooked.
- It doesn't remove trim and baseboard as effectively as other tools.
- It can be a real hazard if not carried or laid down properly. The six-inch tines can penetrate right through a human foot.

ARROW HOOK

Standard Uses

The arrow hook is very similar in design to the New York pike pole. The obelisk-shaped pike has been modified into a very pointed and sharpened arrow shape. There are definitive troughs in the iron head on all four sides of the pike, as well as the hook. These troughs actually help to make strong cutting surfaces on all parts of the tool head and allow it to penetrate almost all building materials except metal. The pike ends in a flattened chisel point. The hook curves off the main pike point, but not severely. The tool head is a socket type.

The shape of the head makes this tool excellent for overhaul. It is not overly heavy (when compared with a national pike pole it's heavy, but it's not as heavy or massive as the New York pike pole). The tool is well designed and very well balanced when installed properly. Its chiseled arrow point and size of the fat hook make it a perfect implement of destruction for overhauling a room. The modified arrow-shaped head makes it a good tool to remove baseboards, door trims, and moldings. The pike will penetrate plaster and lath easily, and the hook is curved slightly so that it will grab a large amount of material and pull it easily on the downstroke.

Gypsum board and lighter material will also fall easily when you use the arrow pike pole. The tool head size will grab chunks of gypsum board and make large holes as compared with the national pike pole, which has a tendency to slice through. The arrow pike pole is very effective for pushing materials as well as pulling.

The head is very similar in design to the New York pike pole and makes the arrow hook as effective as the New York pole for dropping ladders on fire escapes, as described on page 72. The arrow pike pole is also an excellent choice as a roof hook. Its curved hook is able to grab and hold roofing material when opening vent holes. Because its head is hooked, you should insert the butt end of the pole to push down ceilings.

Special Uses

It's just a pike pole. This tool is highly recommended for use in areas of older construction that consists of plaster and lath, wire lath, and heavier types of construction.

Limitations

- Although well balanced, the tool head is heavy, which is always detrimental when mounted on a long pole.
- The tool head needs to be regularly sharpened so that it will penetrate heavy materials adequately.
- The pike and hook are quite dangerous.

ODD HOOKS

There are a small number of variations on the standard firefighting hooks available from different tool manufacturers. Many of them are really odd looking but seem to be effective tools. Not much information about them was available. Only a limited number of departments across the country have such tools, making information on correct and safe procedures difficult to gather. Included here are two tools on which I was able to gather some information.

GATORBACK HOOK

Standard Uses

The gatorback hook is a really strange tool. It has a standard national pike pole head and 10 sharpened teeth running along the top or pike side. The edges of the pike and the hook have been beveled to increase their penetrating ability on the upstroke of the tool.

The firefighter has the option of either pulling the tool like a standard pike

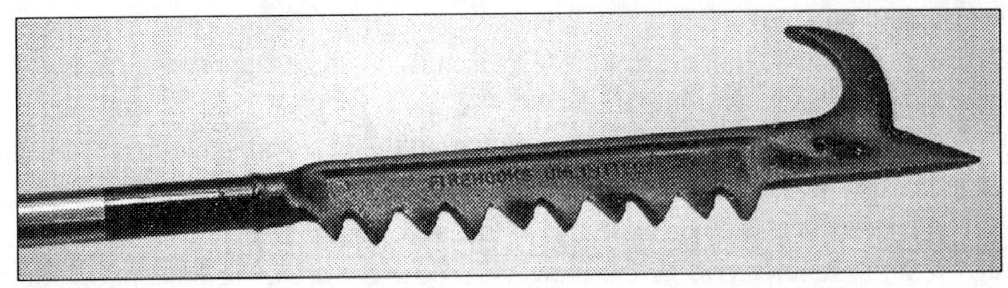
Gatorback hook.

pole or cutting the material by flipping the tool over and using the 10 teeth. The total length of the head is approximately 18 inches.

I haven't seen a tremendous number of these tools in use. The few that I have seen have been four-foot closet models. The firefighters who have used them say they work all right but that the gator teeth usually end up snagging on something and that the pole has to be used like any other pike pole.

Special Uses

I don't honestly know whether there are any special uses for this tool other than what the manufacturer claims it will do.

In-House Modifications

Unknown.

Limitations

- The tool can't always be buried deep enough into void spaces to make use of the gator teeth.
- The tool snags on wiring, conduit, plumbing, drop ceiling fasteners, X bracing, and other materials hidden in void spaces.
- The tool is sharp and difficult to move safely in a structure.
- The tool head is thin and will not grab large amounts of material.
- The hook is very sharply curved, requiring that the head be driven in very deeply for the hook to grab material.
- It is more difficult to use this tool during overhaul due to the shape of the hook.
- It is difficult to use this tool in structures that have limited void spaces behind walls and ceilings due to the size and shape of the tool head.

DRAGONSLAYER™

Standard Uses

The Dragonslayer™ is not really a true pike pole. It is a tool designed to rip and cut gypsum board as well as plaster and lath. It will pull material as you

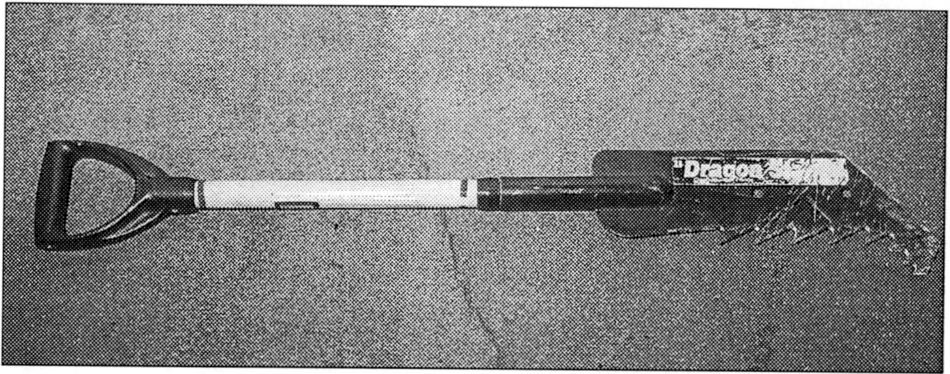

Dragonslayer™.

work with it, but there is no pulling surface incorporated into the tool.

This tool is used like a tree-trimming saw to cut open plaster and lath or gypsum board ceilings and walls. It has a small, triangular point at the tip of the head. Use the point to enter whatever material you are going to cut. The poke method would be the best to use, since the throw method is hazardous with a head this size in a limited-space area. The safest method of using this tool is to open a hole with another tool; otherwise, invert the Dragonslayer and poke a hole into the chosen area using the end of the tool. After the small opening is made, the Dragonslayer can be inserted and used like a saw with no trouble. The tool is supposed to cut on both the upstroke and the downstroke but actually cuts best on the downstroke. Cut until you have the desired hole. The tool is very effective on wire lath.

Special Uses

I could not find any special uses for this tool.

In-House Modifications

No modifications could be found other than good maintenance to maintain the teeth.

Limitations

- The tool can't always be buried deep enough into void spaces to make use of all the teeth.
- It snags on wiring and other materials hidden in void spaces.
- The head is sharp and a threat to carry inside.
- The head is thin and must be worked hard to cut.
- There is no hook.
- The shape of the head makes it more difficult to use during overhaul.
- The size and shape of the tool head make it impossible to use in structures lacking adequate void spaces behind the walls and ceilings.

CHAPTER 6: PERSONAL TOOLS

In many instances on the fireground, the availability of a tool may make the difference between effecting a rescue and doing a body recovery, between getting yourself out of a tight situation and not. Firefighters, especially officers, should always have a tool with them.

A lot of the standard firefighting hand tools are sometimes too clumsy to take with you when stretching handlines or performing some of the other tasks you are required to do. There is still no excuse for not having a tool available.

In this chapter, we'll look at tools that are downsized versions of standard firefighting hand tools. They are easily carried and efficient when used correctly. Keep in mind, however, that these tools do not replace the standard firefighting tools—they supplement them. Many of the personal tools are severely limited in capability, especially the prying tools. You can count on them to perform well if you know how to use them properly and you fully understand their limitations. Personal tools cannot replace the full-size firefighting tools.

OFFICER'S HALLIGAN HOOK

Standard Uses

The officer's halligan hook is a full-size tool head on a shorter metal shaft. One side of the head is completely flat, which makes it easy to slide along the floor. The metal shaft is covered in foam, which makes a good, firm grip. On the end of the tool is a gas shutoff.

This is an excellent tool for an officer to carry. Its flat side will slide along the floor while you crawl, and its length will add reach to your search. It's an excellent tool for sweeping under furniture, into cabinets, or under beds when searching for children.

The halligan hook head gives you all the prying options available on the full-size pole. The only thing missing is the amount of leverage you will have for big prying jobs. Limit the use of this tool for making inspection holes or for prying back trim and baseboards to check for fire extension.

The smaller size of the tool has advantages over the longer pole version. You can swing this tool more effectively to chop open holes in walls. Giant swings are unnecessary—short, strong blows will open a decent hole. Insert the tool head and pull as you would a standard-size pike pole. An inspection hole will be opened rapidly.

The officer's halligan hook is an extremely useful tool for engine company officers. It can easily be carried in a hammer holster such as the kind carpenters use. Attach the holster to a truck belt and the tool will always be with you. Truck work that must be performed immediately, with the exception of opening ceilings, can be performed by the officer quickly, and the fire can be controlled until truckies can be brought in to completely open up the void spaces. Windows can be vented with the tool, saving possible injury to firefighters who might otherwise use a helmet or body part to break glass. The tool will also efficiently remove all the window sashes, creating an egress point for emergencies.

The gas shutoff on the butt end allows the officer to control gas utilities quickly while performing his size-up walk-around. It will also allow him to pry open windows and break window and door glass for quick forcible entry.

This tool can't replace the full-size halligan hook pole, but it will make the engine company more efficient without hampering their mobility.

Special Uses

There are no real special uses for the officer's halligan hook. Read the section on the halligan hook in Chapter 5 for many of the uses of this style of tool head.

In-House Modifications

There are two in-house modifications recommended. The first is to keep the beveled edges of the tool relatively sharp. However, because you will be crawling around in dark, confined spaces with other firefighters, keep the edges sufficient to pry with but dull enough so as not to cause injury. The second modification would be to repaint the tool or mark it with reflective markings. The tool is black when it comes from the manufacturer and is hard to see.

Limitations

- The tool is short, so it lacks some leverage compared with a full-size halligan hook.
- Swinging it in the dark and smoke can cause severe injury. When breaking glass, don't overswing. The tool is weighty enough. Overswinging may pull your hand through the window, causing severe cuts.
- The tool comes from the manufacturer painted black. It's hard to see if you drop it or put it down while searching.
- This tool has limited forcible entry capabilities.
- The tool is much too short for pulling ceilings.

OFFICER'S TOOL (AKA O TOOL)

Standard Uses

The officer's tool is a multipurpose forcible entry and search tool, designed and specified by FDNY. The original tool was invented by Captain Bob Farrell, Ladder 31, FDNY. It is 17 inches long, is made of high-quality steel, and weighs 2½ pounds. The tool is basically a block-type head welded onto a steel shaft. The back side of the block, better referred to as the adze, is a striking surface. The adze is slightly curved, and a lock-pulling tool known as the A tool is machined into the front side of the block. The A tool is a beveled cutting surface very similar to an ordinary nail puller. The A tool is wide enough to allow the face of a lock cylinder to slide into it so that it can be pulled off. The beveled edges are on the inside of the tool head, and the outside surface is smooth to allow the tool to be driven in and behind lock cylinders.

The tool shaft is usually covered in foam, which may or may not have little pockets to accommodate a through-the-lock forcible entry key tool. The bottom side of the tool has a slightly angled prying surface; the angle of the surface is opposite that of the A tool in the head.

The officer's tool functions well as a light pry bar. It will pop open standard lightweight residential doors and light commercial doors such as those found in office buildings. It is used in virtually the same manner as a claw tool, as described in Chapter 3. This tool is 10 inches shorter than a standard prying bar, however, and doesn't have the same leverage capabilities. Just don't overexert pressure on this tool.

The officer's tool can be used to open windows, as long as you keep the bevel of the prying end well dressed. It's really designed to function as the primary tool for through-the-lock forcible entry. The A tool machined into the adze end will effectively pull most lock cylinders from the door, allowing you to gain access to the lock mechanism to trip it with the key tool.

To remove a lock cylinder, position the A tool on top of the cylinder at a 45-degree angle. Strike the top, driving the beveled edges down and behind the cylinder. Lift up on the tool as you strike to allow it to get in and behind the

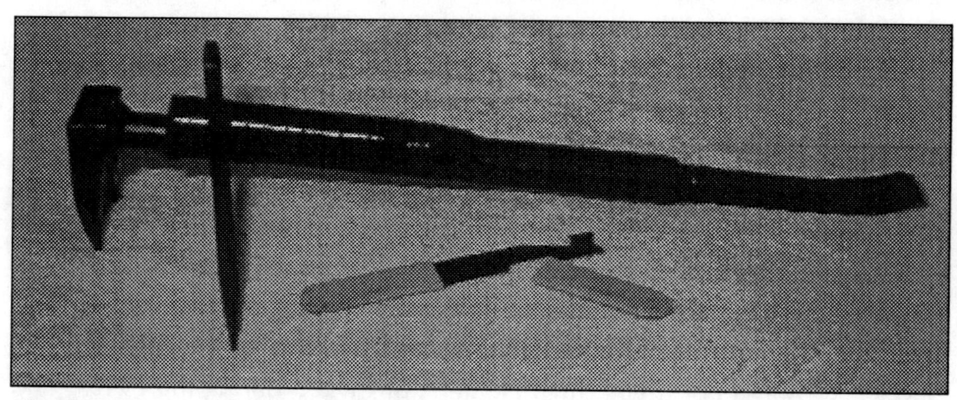

Officer's tool.

lock face and to grab hold. Strike the tool until you are sure that it has a firm hold on the lock cylinder. Pry up. Always pry up. This minimizes the chances of damaging the lock mechanism, making it inoperable. The lock cylinder should pull out. Some locks have case-hardened screws, so you may be there awhile, struggling. Keep driving the tool as needed. Eventually the lock will pull out. Now the mechanism is exposed for you to trip it with the key tool.

The officer's tool extends your reach during searches and will allow you to vent windows as you go. It will also create inspection holes, pull baseboards and trim, and perform other limited overhaul jobs. This tool is extremely versatile and should be available to anyone riding in the right front seat of responding apparatus.

Special Uses

The entire premise of this tool is that it is a special-use tool. There are no other specific uses that could be categorized as special.

In-House Modifications

Paint the tool a color you can see. Otherwise, it is ready for use when you get it from the manufacturer. Keep the bevels sharp. You may want to increase the sharpness of the bevel on the prying end of the tool since it sometimes is a little fat to get a good initial purchase when prying.

Limitations

- It is not a full-size bar, and its total leverage capabilities are limited. It cannot and will not force medium to heavily secured doors—a standard pry bar will be needed. The tool can be bent or broken if too much force is applied.
- The two prongs at the end of the A tool are extremely sharp and can cause severe puncture wounds. Always crawl with the points down!
- The tool is small and black. It is hard to see.
- Its limited leverage consequently limits its suitability for overhaul.
- For pulling ceilings or heavy overhauling, use a full-size tool.

MINI-HALLIGAN BAR

Standard Uses

A personal halligan bar (full-size) is the best prying tool available to the firefighter. The mini-halligan bar is a modification of the full-size bar developed by Deputy Chief Hugh Halligan of FDNY during the 1940s. The bar is much lighter, weighing only 4½ pounds. It is shorter than the standard bar by 10 inches. The adze on the mini-halligan bar has been replaced with an A tool head, similar to that of the officer's tool. A narrower, slightly angled hook or pike extends from the A tool head.

This section of the chapter will describe the uses of the mini-halligan bar, which is preferred by firefighters as a personal tool. It must be stressed that this tool does not replace the standard-size halligan bar. It is used as a personal tool by firefighters who understand that a full set of irons is close by to back them up if they need it. None of the tools in this chapter should be construed as replacements for their heavier, full-size counterparts.

The use of the mini-halligan bar is still a matter of leverage. With the smaller bar, obviously, you have less. The design of the tool allows for multiple functions with one tool, but leverage is the key. Mini-halligans are 20 inches in length and weigh about four to 4½ pounds. These tools aren't usually a single piece of forged steel—normally the tool heads are welded onto a steel shaft. Often these tools have tubular shafts or bars, which are substantially weaker than a solid piece of steel.

The adze end of the tool has the contours of a standard halligan bar. The A tool is machined into the adze to create a formidable through-the-lock forcible entry tool. The shaft is usually made of aircraft steel. Some tools have a diamond-type grip surface machined into them; others have the foam-style covering for added grip. The fork is not broad and is relatively straight with a slight curve. On most tools, the fork is four inches long and tapers into two straight, beveled tines (the bevels are on top of the tines). The spacing between the tines allows gas valves; small, nonhardened padlock hasps; and other objects to be levered by the tool. The bottom side of the fork is called the beveled side, and the top side of the fork or "dished" side is the concave side. There are no striking surfaces on the fork. This smaller 20-inch version of the halligan will perform numerous functions when properly used. To start, it is a good bar to use in conventional forcible entry.

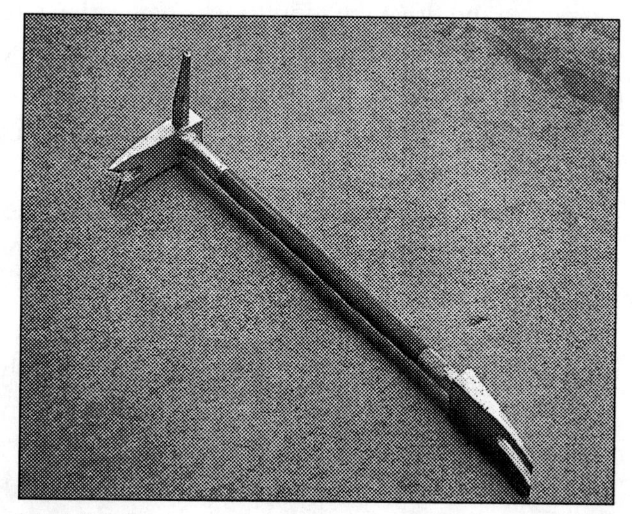

Mini-halligan bar.

For inward-swinging doors, two firefighters can make quick work of a lightly to moderately secured door. Place the bevel side of the fork against the door, about six inches above or below the lock. If it is a steel or metal-clad door, go get the bigger halligan! Angle the bar slightly toward the floor or ceiling. Tap the mini-halligan with the back of a flathead axe or, better yet, a mallet or smaller hammer. An eight-pound axe slamming into this tool may break the weld. This tool should not be forced into areas that call for a heavy-duty, full-size pry bar. It is not designed for heavy doors, high-security locks, or steel doors. It will work fine in residences and light-commercial buildings.

Another method for inward-swinging doors is to drive the hook of the small halligan completely into the doorjamb six inches above or below the lock. Push the mini-halligan bar down and the door will open. It is very important to push down. In trying all the different methods, I discovered that you work twice as hard physically when you pry upward, and the door usually doesn't open! Use leverage to your advantage, especially in this case, since the mini-halligan bar is only 20 inches long. A moderately heavy tubular dead bolt may stop you completely.

Outward-swinging doors are another story. Flush-fitting doors can be forced using either the adze end or the fork end of the mini-halligan. To perform the fork end technique, which is very effective, place the concave side of the fork toward the door, cant the tool slightly toward the floor or ceiling to get a better purchase, and drive the tool using a flathead axe or other striking tool. As the mini-halligan is driven in, move the tool perpendicular to the door to prevent driving the fork into the jamb. When the tool has spread the door as far as possible, force the adze end of the mini-halligan away from the door and it will open.

Another technique is to use the adze end of the mini-halligan bar. Put the adze of the tool six inches above or below the lock. Drive the adze into the space between the door and the jamb. Drive the adze deep. The bar should stand out by itself perpendicular to the door when you're finished driving it in. When you are sure that the adze is sufficiently into the space, pry down and outward with the fork end of the mini-halligan to force the door open.

If you run across recessed doors or a door with a wall next to the lock, use the adze end of the tool. Place the adze end of the mini-halligan six inches above or below the lock and drive the adze into place. Pry downward and out with the forked end of the tool to force the door open. This operation is a bit clumsy, so be extremely careful when driving the tool into place. One firefighter should do it.

The mini-halligan bar is excellent for prying open windows. Insert the fork of the tool under the bottom sash of a double-hung window and tap the tool into place. Pry down, and the screws of the window lock should pull out.

This mini-halligan will also work as an A tool when confronted by locks. Use it in the same manner as you would the officer's tool, described on pages 101-102.

Special Uses

There are no real special uses for the mini-halligan bar. It is a personal-use tool, and it cannot be stressed enough that this implement does not replace a full-size bar. The tool functions well for light overhaul, for making inspection holes, and for light forcible entry.

In-House Modifications

The tool is black and hard to see. Add color or reflective tape so you don't lose it. Some tools have the foam grip; some do not. On those tools that don't have the foam grip, you might consider adding French hitching. The grip should be at least two inches from the fork end of the bar, up to two inches below the adze.

Limitations

- The tool is too short. Twenty inches does not give you a lot of the leverage you need for forcible entry.
- The tool is welded, not forged.
- The points on the end of the A tool are very sharp and can cause serious injury. Always crawl with the points down.

TRUCKMAN'S TOOL (AKA TRUCKIE TOOL)

Standard Uses

The truckman's tool came about due to the shortcomings of the smaller versions of the halligan bar. Leverage is a key factor in using any tool, and the smaller versions of the halligan just won't do all of the same functions as the big bar. The truckman's tool is not a replacement for the halligan, either, but it does have two major improvements over its smaller cousin.

The first improvement is that the truckman's tool has a full-size adze head like a full-size bar. The adze end of the tool gently curves and flares out slightly from the tool shaft to the end. The adze has a long and wide A tool machined into it. This A tool is wider than that of the officer's tool or the small halligan. It will receive some of the larger tubular-style locks much more easily than the other two types of tools. Secondly, the head of the tool has three striking surfaces: the top of the tool—that is, on top of the adze—and on both sides. These surfaces have been designed to receive relatively heavy blows from a striking tool.

The tool is 16 inches long. On the other end is a full-size curved chisel prying surface similar in design and function to the chisel prying tool on the

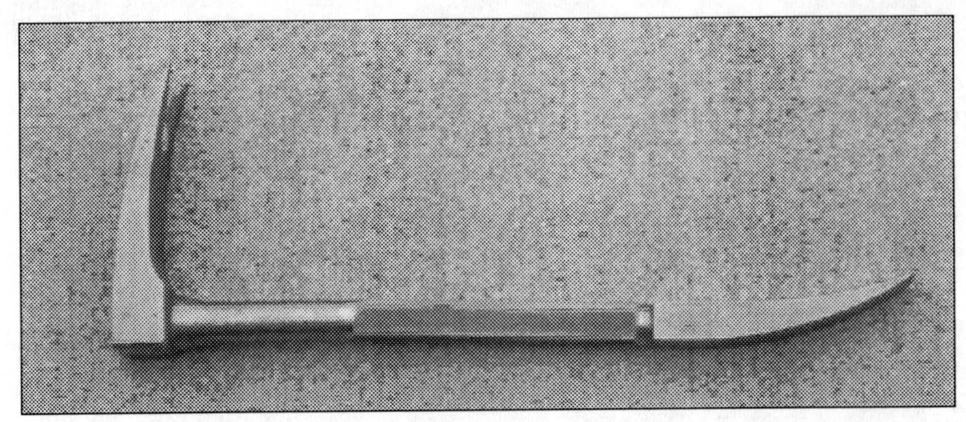

Truckman's tool.

end of a roofman's hook. This prying surface allows you to pry off hasps, fasteners, latches, and other locking devices. The prying end can be driven in, but with only 16 inches of lever, don't go too deep. The tool will be a big nail, and you won't be able to get it out. On some versions, there is a forked end like the small halligan. The fork is at a 90-degree angle to the adze, much like the Chicago patrol bar. The fork can be used in the same way as the forks on the small halligans. The fork is not full-size, so its ability to pry is limited.

Through personal use, I have found this tool to be a better A tool than the mini-halligan or the officer's tool. The A tool's full-size head and wider opening make it very effective on different types of locks. It does lack the prying capacity of a small halligan or O tool, however. The tool is welded like the small halligan and the officer's tool, so it can break if misused or poorly maintained. It lacks a hook or point like the small halligan, but that's no great loss. It isn't supposed to replace the halligan—just augment it.

Special Uses

The truckman's tool has no real special uses. It is a special-use tool unto itself.

In-House Modifications

There aren't really any modifications that can be made to this tool to improve its performance. It is made of shiny steel, so other than a personal mark or two, you don't need to paint it. The chisel end can be filed to a finer taper and bevel to be used for jimmying open light residential and commercial doors. The version with the forked end should also be filed to improve the bevel of the fork tines. The machined knurled grip in the handle provides a sufficient grip surface. French hitching would be a lot of work for this small tool, and it wouldn't really improve the grip much.

Limitations

- The tool is short, and doesn't have a great deal of leverage. When the tool is driven in with a striking tool, it can be difficult to get it to move.
- The points on the end of the A tool are very sharp. Keep them pointed downward when crawling.

FENCER'S PLIERS

Standard Uses

Fencer's pliers are commercially available items. They were first brought to my attention many years ago by a fellow Davenport, Iowa, firefighter. He carried a pair in his coat pocket and used them at almost every fire we went to. The pliers were again brought to my attention by Firefighter John Grasso of Engine 48, FDNY. John also carries a pair in his coat pocket.

These pliers are odd-looking things. They are used in rural farm settings for setting up and repairing fencing. They have a pick point, wire nippers, wire cutters, a pliers jaw, a small striking face, and insulated handles.

The point end can be used to break window glass. The pliers are strong enough and provide enough leverage to turn the gas valves on most residential gas meters. They are excellent for removing battery cables from their posts. The small hammer can be used to tack plaster and lath in place during salvage operations.

They are also capable of cutting chain-link fence wire, metal staples, and de-energized wires. They are great for grabbing that elusive hood latch cable during a car fire.

All in all, these pliers will give you the most for your dollar. They're a handy tool to be carried in your coat pocket or a great addition to the special tools kit.

CHAPTER 7: SEVERAL-IN-ONE TOOLS

This chapter deals with some pretty controversial items. Throughout fire service history, tools have been designed to try to perform multiple functions. Most have long since been retired, ushered out of the fire station on the same piece of apparatus they came in on—many of them with their original coat of paint still intact.

The tools used in the fire service in this country date back more than 390 years to the first permanent English settlement at Jamestown in 1607. Some of the tools we use today stem from that settlement. Most notable among them are the pickhead axe and the pike pole. A lot of tools have come and gone since then; we always seem to drift back to some very basic, durable tools.

There are four several-in-one tools this chapter will deal with. There may be more, but four is plenty. We'll start with the most recent tool and work toward the older ones.

T-N-T TOOL, FORMERLY THE DENVER TOOL

Standard Uses

The T-N-T tool is a combination of four tools in one: sledgehammer, axe, pike pole, and pry bar. It can be used to ventilate, pull ceilings, and perform conventional forcible entry and overhaul.

This tool is available from its manufacturer in three different lengths and four different weights. The standard model is 40 inches long with an 8½-pound tool head. Its overall weight is 13½ pounds. Other weights and sizes stretch from 30 inches long with a 6½-pound head to a 40-inch tool with a 6½-pound head. Total weights range from 10½ pounds to 13½ pounds.

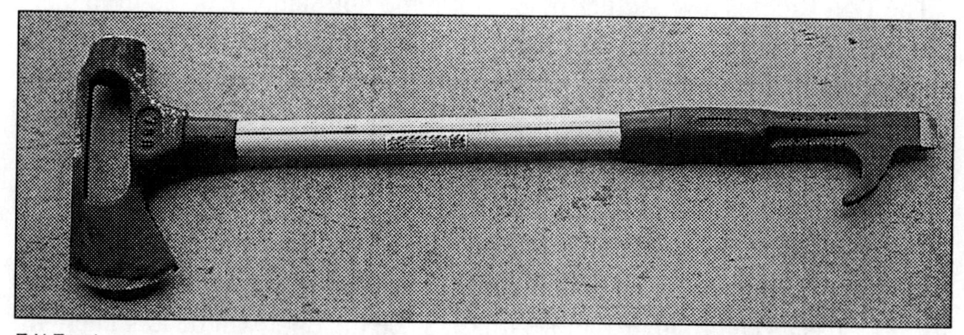

T-N-T tool.

The tool resembles an axe. The working head has a striking surface on one side and a broad-bit axe on the opposite face. There is a hole in the tool head that functions as a handgrip. The fiberglass handle is stout and ends in a combination chisel/pry point and pike pole head. The chisel point is wide and flat and very sharp. The hook is thin and flat.

Using the tool is much like using an axe; it requires the same attention to stance and swing. To use the T-N-T tool as a pike pole is a chore. At a maximum length of only 40 inches, it will be difficult to reach up over your head to pull ceilings. However, the tool does work well when pulling walls. To use it as a pike pole, grab it through the hole in the tool head. Keep the blade side turned away from your face. Use the poke method for getting the tool into the ceiling. The chisel point will pierce plaster and lath with relative ease. Don't get carried away trying to insert the tool, though. There is an 8½-pound sledge head with a cutting blade real close to your face. It is unforgiving if you hit yourself.

The narrow pike hook works identically to the national pike pole. It makes small tears in the gypsum board and cannot grab a lot of material on the downstroke. Use the tool as a pike pole sparingly. Working with it over your head is very tiring.

The T-N-T tool can be used to breach walls. Use the axe to cut through walls efficiently, or flip it over and use the striking end to smash through them. During overhaul, the T-N-T tool can be an effective tool to open walls, floors, and other surfaces. The blade is a very effective tool for sliding behind baseboards, door trim, and moldings. The amount of pressure you will need to exert to remove such materials will depend on how well it was installed, but generally you will be able to get it off with the T-N-T tool. Remember, you are applying unnatural force to the handle when prying. If whatever you are trying to remove resists, move the tool to a new purchase point. Overapplying pressure may snap the handle.

To remove window moldings, door trim, and other materials, drive the chisel point/prying end down and behind the molding, then pull the tool toward you. To remove baseboards, use the chisel point.

The prying end is also an excellent tool for removing floorboards. Drive the chisel into a groove between floorboards and push the tool down, levering the floorboard up. The objective here is to use the tool as a long-handled chisel. Make sure that you have a strong purchase in the wood before you begin prying.

The T-N-T tool is a very efficient tool for removing standard doors from their frames. In most residential and light commercial buildings, doors are only attached to their frames by hinges screwed into wooden studs. To remove a door with little or no damage to either the door or frame, insert the blade of the axe between the door and frame either just above or just below the top hinge. Shut the door. Exert slight pressure on the handle and pry the door away from the frame. The screws will pull out. Open the door, turn the tool upside down, and repeat the process at the bottom hinge. The door can easily be removed. If the door has a center hinge, remove it last, since the door will be less likely to get stuck that way. This is an excellent method for getting doors out of the way during firefighting operations and overhaul. It is also an easy method for obtaining something solid (the door) to act as planking over a hole in the floor. Little damage will be inflicted on either the door or door frame.

If you have well-maintained tools, the handle of your T-N-T tool is also effective for removing lots of plaster and lath in a short time. While overhauling, open up a decent-size hole three to five feet off the floor. Insert the handle of the tool down into the bay. Grasp the blade and pick and pull toward you. The handle will pull large amounts of plaster and lath.

Special Uses

The T-N-T tool is capable of performing several tasks in auto extrication, including making holes to act as purchase points for heavier hydraulic tools. It will also remove auto glass easily.

In-House Modifications

Don't modify this tool.

Limitations

The T-N-T tool has several limitations:

- The tool is sharp at both ends. When using it, there is always a sharp end in motion.
- When working overhead, the firefighter is pulling a heavy hammerhead close to his own head.
- The tool is too short to be an effective pike pole. Working overhead with this tool is very tiring.
- It lacks any real curved surface for prying. There is a chisel end, but no adze.
- There are no real defined fulcrum points for prying.
- Using the hole in the head for a handle may cause wrist damage if the pike pole end of the tool hits unexpected hard materials.
- This tool can't be easily repaired in the firehouse.

CINCINNATI TOOL

Standard Uses

The Cincinnati tool was developed by a lieutenant in the Cincinnati Fire Department. It is very similar to the T-N-T tool in the types of tools that it has combined. The Cincinnati tool has a pickhead axe, a prying surface, a nail puller, a gas shutoff, and a pike pole all in one. The tool is one-piece construction, all of stainless steel. It comes in one length, 26 inches, and weighs a very light 10 pounds.

The axe head end has a cutting-type axe surface on one side and a pick tool on the other. There is a hole in the head, big enough for a gloved hand, so that the tool can be used as a prying or pike pole. The other end of the tool is a very odd-looking hook and nail-pulling tool. The hook is almost exactly like the hook of a national pike pole. It is flat, sharply curved, and narrow. On the bottom side of the hook is a nail puller/forked prying end. The prying end is very narrow, and it forks into two narrow tines. The prying tool is curved and functions just like the claw on a claw hammer. There is a gas shutoff machined into the metal area between the hook and the prying surface.

The technique for using the tool is exactly the same as the T-N-T tool with one exception: There really is no prying surface on the Cincinnati tool. There is a narrow nail puller, with no real ability to pull large amounts of material. There is no chisel or adze.

Special Uses

There are no special uses for this tool.

In-House Modifications

Do not modify this tool.

Limitations

Because of its similarity to the T-N-T tool, its limitations are similar.

- The tool is sharp at both ends. When using it, there is always a sharp end in motion.
- When working overhead, the firefighter is pulling an axe head close to his face.
- The tool is too short to be an effective pike pole. Working overhead with this tool is very tiring.
- It lacks any real usable surface for prying. There is a no chisel end or adze.
- The fulcrum points are inefficient for prying.
- Using the hole in the head for a handle may cause wrist damage if the pike pole end of the tool hits unexpected hard materials.
- This tool can't be easily repaired in the firehouse.

PRY AXE

Standard Uses

The pry axe is a multipurpose forcible entry and search tool. It has been a

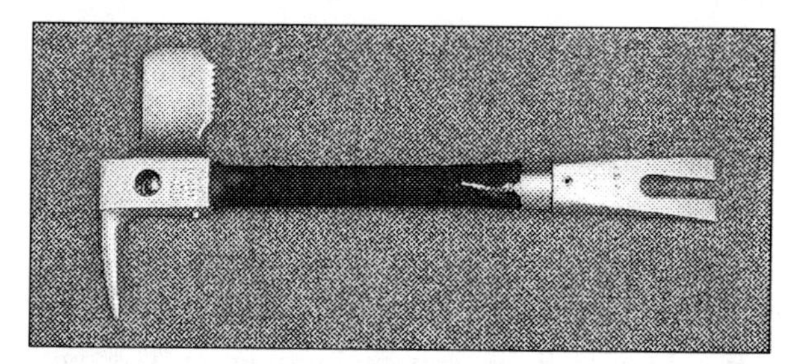

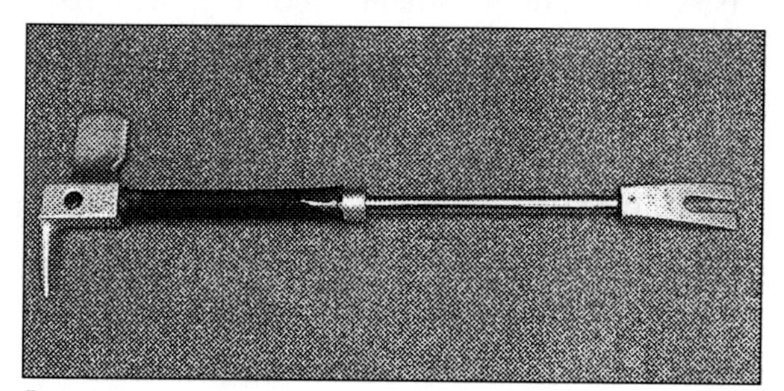

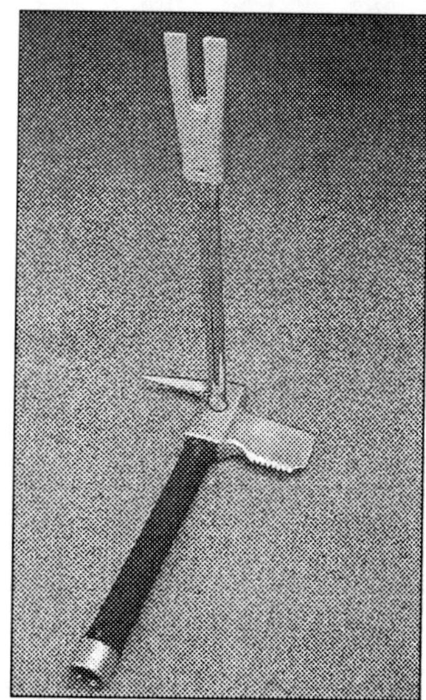

Pry axe.

standard firefighting tool for many years. The functions of the pry axe include cutting, prying, utility shutoff, forcible entry, and glass removal. The tool is about 18 inches long and weighs approximately four pounds. It is actually in two pieces—there is a removable steel bar that slides out to make an additional prying handle. It has a block-type head that is welded onto a steel shaft. One side of the block has a cutting blade with serrated teeth located on its bottom side. Opposite the cutting blade is a tapered pick. The tool shaft is covered in rubber. At the bottom of the tool is a fat fork, which is attached to a steel rod that slides up into the handle of the pry axe. There is a release button on the pick side of the tool head to release the bar. If you twist the fork a quarter turn, the bar will be released entirely from the tool. The fork also functions as a gas shutoff.

The tool can function like a small axe, capable of cutting material with the blade or smashing it with the pick. It's a bit lightweight, though, and the head and cutting surfaces are easily broken or chipped.

The pry axe functions well as a light pry tool. It will pop open standard lightweight residential doors and light commercial doors like those found in office buildings. There are several techniques for using the pry axe. To open an inward-swinging door, use the axe part of the tool like a slap hammer to drive the fork into the door. This is a small tool with small surfaces, so don't drive the tool in too deep and overload it. As the fork end is struck, slowly move the bar perpendicular to the door to prevent the fork end from penetrating the interior doorjamb. Exert pressure toward the door, forcing it open. Don't overexert pressure on the tool; it doesn't have the leverage capabilities of bigger, heavier tools. Use the slap-hammer technique for doors that open outward as well.

To open windows with the pry axe, keep the bevel of the fork end well dressed. Insert the fork end of the tool between the bottom rail of the window and the sill. Once set, pry down on the bar. The screws holding the window lock should pull out. The axe blade can also be used to open windows. To perform this, first release the bar from the handle of the tool by twisting the fork a quarter turn and sliding it out. Insert the end of the bar into the socket hole in the side of the tool head. Slip the blade of the axe under the window frame and push or pull the steel bar. The window lock will pop.

During overhaul, the pry axe can be an effective tool to open walls. The blade is a good tool for sliding behind baseboards, door trim, and moldings. The pick on the pick head will not allow you to pry the baseboards easily because it's too small. To use the blade to remove baseboards, place the blade behind the baseboard at the point where the baseboard meets the wall. Once the blade is well set, pull the handle toward you. If the blade has trouble getting into the space, release the fork from the tool and tap the blade into place. Reinsert the fork into the handle, but leave it sticking out at maximum extension, which will increase your leverage and allow you to get the baseboard off.

The pry axe extends your reach during searches and will allow you to vent windows as you go. It will also create inspection holes, pull baseboards and trim, and perform other limited overhaul techniques

Special Uses

The entire premise of the tool is for special use. The pry axe is not used much anymore. It has been replaced by newer and stronger tools. Many of the firefighters who really knew how to use this tool have long since retired.

In-House Modifications

No modifications should be made to this tool.

Limitations

- It is less than full-size and therefore has less leverage capability. It can't even breach medium-security doors and will bend or break if too much force is applied.
- It is small and easy to lose.
- Its limited leverage makes it suitable for small overhaul jobs only.

HUX BAR

The hux bar is a hydrant wrench gone crazy! This tool is in every old manual that I've run across, but there is never any explanation as to how to use it. The tool itself is still around, and fire departments all across the country have them on active engine companies.

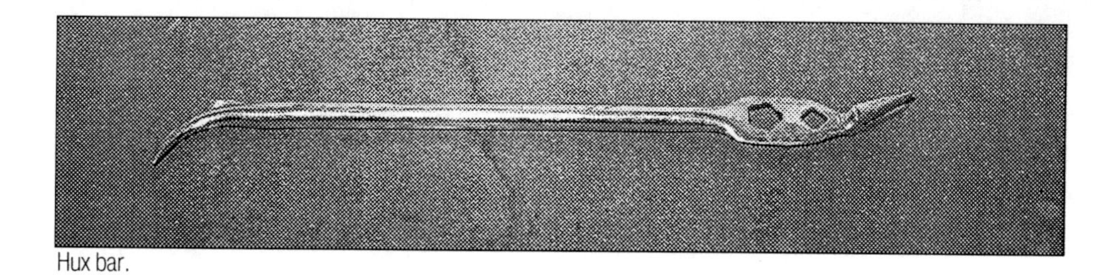

Hux bar.

Standard Uses

The hux bar is first and foremost a hydrant wrench, and a poor one at that. The tool is chromed tubular steel, with a very thin, flat hydrant tool at one end and a curved, tapered prying point with a small fulcrum at the other. The hydrant wrench end has two openings for hydrant operating nuts. One is square; the other is pentagonal. Most of the tools currently in use have had these openings severely rounded out by trying to open tough hydrants. Attached to the flat hydrant tool end is a two-pronged nail puller.

I cannot in good faith say anything positive about this tool. I've asked many firefighters who have used it, and most of them have admitted to breaking or bending it when they encounter tough hydrants or a medium-security door or window. I've seen intact tools in fire departments, but the tool has been taken out of service and is in the storeroom or polished and displayed on the parade rig. Most of the tools that I have seen still carried on active rigs are rusted and broken. The nail puller is most often bent or broken off completely. The hydrant wrench holes are rounded out, and the tapered prying point at the end has been straightened.

If you have this tool, do it and yourself a favor. Take it out of service. It has survived long past its usefulness and is not a tool on which you want to stake your life.

Special Uses

There are no special uses. This tool cannot perform basic tasks.

In-House Modifications

Polish it and put it in a display case or on the antique rig.

CHAPTER 8: SPECIAL-PURPOSE TOOLS

Not every situation you will face on the fireground can be handled by standard fire service tools. Over the years, firefighters across the country have been faced with special situations where they needed a certain type of tool to accomplish an assignment quickly and effectively. These special situations can range from special types of lock assemblies to high-security padlocks to brick walls and a whole range of other problems encountered on the fireground. The standard fire department tools may not be sufficient to get the job done. Firefighters, being who they are, designed and built those special tools to provide a more efficient and professional service to the public.

Often the tools themselves were borrowed from the seedier sides of life. Burglar tools, jimmies, and slip knives were not the standard tools carried by the local hardware store or reputable toolmaker. There were no instruction manuals included in the package. Over many years and hundreds of thousands of situations, special-purpose tools became standard fire department inventory.

Every firefighter reading this book must remember one thing about these tools: Don't get caught carrying them without identification! These are very effective burglar tools and will raise questions about your intent if you are caught off duty with a bag of them in the trunk of your car and no identification. As you acquire these tools either for personal use or as department issue, mark them well, and don't let them fall into the wrong hands. It can be very embarrassing!

This chapter will examine several different types of special-purpose tools. Many of these tools are what are known as limited-use tools. Limited-use tools perform very specific functions and cannot and should not be used for any other purposes. Selecting the appropriate tool for the assigned task is a basic and important skill for every firefighter.

BAM-BAM TOOL

Standard Uses

The bam-bam tool is simply an automotive dent puller doing fire duty. This tool was borrowed from the car theft industry and has practical uses on the fireground and on the scene of vehicle accidents as a lock puller.

Before running out to the local auto parts store and getting a dent puller, read the rest of this section. Bam-bam tools are not all alike, and to get the most efficient use of this limited tool, you'll need to get the right one.

The bam-bam tool should be of heavy steel, with a sliding slap hammer of sufficient weight so that a more than adequate force is applied with each movement of the slide. The shaft of the tool must be heavy steel, and the handle should be threaded onto the shaft with a beefy stop to take the beating from the sliding hammer. Lightweight-type dent pullers for home auto repair are really inefficient for fire department use. The types of locks you will be pulling require a tremendous force to get them out of the lock body. Lock manufacturers design locks that will defeat the use of this tool. Keep that in mind when sizing up the situation. The bam-bam may not be the right tool.

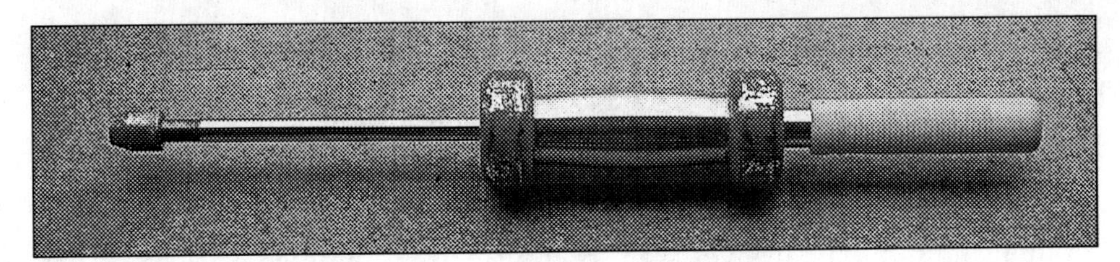

Bam-bam tool.

The last but most important aspect of the bam-bam tool is its working end. There should be a threaded collar attachment that will allow you to change the screws easily. Case-hardened sheet metal screws are attached and held tightly by this collar assembly. Some specially designed fire service bam-bam tools also have special threaded spike attachments. For heavy-duty locks, the spike works better for pulling lock cylinders.

Size-up of the lock assembly is critical for deciding whether the bam-bam tool will be effective. A heavy duty bam-bam will have a higher rate of success than lesser ones.

When using the bam-bam tool on padlocks, look at the keyway and the shackles. If the shackles are greater than a quarter-inch in diameter, this is a high-security lock. Padlock manufacturers have added a metal security ring around the brass tumbler assembly to prevent the lock from being pulled from the body. On locks manufactured in the United States, particularly Master Locks and American Locks, this security ring presents a problem. On locks manufactured in foreign countries, this ring is fake and is only designed to make their product look like the high-quality U.S. item.

If you have a lightweight bam-bam tool, you will not pull the cylinder out of the bodies of these locks! A heavy bam-bam is required, and another firefighter should be sent to get the rotary saw with a metal cutting blade just in case.

If the lock is not a high-security type or is imported, use the bam-bam tool. Insert the screw into the keyway, and screw it into the soft brass at least three full turns. Try to get the screw as deeply into the cylinder as possible. Once the screw is set, slide the slap hammer up to the top of the tool (toward the screw end), and then sharply slam the hammer back toward the handle. This action should be sharp and hard. You are trying to rip the lock tumbler out of the body of the lock. The lock manufacturer doesn't want that to happen easily, so slam hard! Three hard slams and the lock cylinder should be out. Once you have pulled the cylinder, insert a screwdriver or key tool and trip the lock mechanism to open the shackles.

If the screw or spike pulls out of the lock before the cylinder comes out, change the screw on the end of the tool and try the procedure again. Always have an adequate supply of case-hardened screws for the bam-bam tool with you. The screw must bite well into the brass of the lock tumbler, or it will simply strip out and not do anything. The screw may or may not bite into the keyway if it is made of steel or some other metal besides brass. Size it up first, and have an alternative method ready!

The bam-bam tool is used to remove lock cylinders.

Mortise and rim-lock cylinders can be pulled the same way using the bam-bam tool. Size-up cannot be stressed enough. High-security locks are tough to begin with and may require a different approach using different tools.

Using the bam-bam tool requires practice. Don't use this tool under emergency conditions when you have to get in someplace and haven't practiced with it. Obtaining old locks, cylinders, and other assorted junk from local locksmiths can be a great help. Pulling cylinders on structures that are being demolished is another way to practice. Cars used for auto extrication practice should head back to the junkyard with all of their locks pulled.

Special Uses

The use of the bam-bam tool is a special use in itself, but this tool can be used in emergency situations to open car doors and car trunks. The use of this tool should be limited to emergency situations. Locksmiths or wrecker operators can get into automobiles more efficiently and with less damage than we can; plus, they will have the liability.

The bam-bam tool screw is inserted into the car door lock or trunk lock and seated firmly. A minimum of three turns to set the screw are required. Once set, slam the slap hammer firmly against the stop. Continue until the lock has been pulled completely out of the car. Using the bam-bam to pull the lock out is preferable to using the point of a halligan bar or some other tool to drive the lock inward. By driving the lock in, the mechanism may be damaged, preventing it from opening. Pulling the lock cylinder out reduces the likelihood of this.

In-House Modifications

There are no in-house modifications that should be made to this tool. By purchasing a heavy-duty model, preferably one designed for fire service use, any additions or work other than routine maintenance will be avoided.

The bam-bam tool is a difficult tool to carry and can easily pinch your fingers. One suggested way of carrying it is to cut a piece of PVC pipe and then to attach a handle and slide-on caps. This makes a good carrying case and allows you to keep a bag full of screws inside. Having the tool in a case keeps shifty eyes around the fireground from quickly identifying it if a compartment door is left open. This is a primary tool for car thieves and burglars, and its security should not be taken lightly.

Limitations

The bam-bam tool is limited by its own nature. It is a special-use tool for the fire service, and as lock manufacturers continually improve on the strength and complexity of locks, this tool will eventually phase itself out. The limitations of the bam-bam tool include:

- This tool will not work on all types of lock cylinders, especially those with case-hardened tumblers or tumbler protectors.
- The bam-bam tool requires a lot of practice for proficiency.
- Improper use of the tool may jam a lock beyond all hope of forcing.
- It's heavy.
- If left out in public view, it will probably be stolen.
- Fingers or hand webbing pinched by the slap hammer will be badly injured.

HOCKEY PUCK LOCK BREAKER

Standard Uses

Hockey puck lock breaker is a long name for a very common tool: It's a big pipe wrench. Wait! Before you rush out and buy a big pipe wrench and rename it, there are some basic things you need to fully understand about this tool!

The hockey puck lock breaker is designed for one function only: A firefighter uses it to apply force to a high-security padlock to break it. This tool will break an American Series 2000 lock. This is just one of the many high-security locks available. The American Series 2000 lock, the hockey puck, so nicknamed for its looks, is hardened steel with no visible shackles and a recessed keyway. It is a little round thing that weighs about 1¾ pounds and measures 2¾ inches in diameter and about 1¾ inches thick. The shackle is recessed in the back of the lock.

The lock is tough. The pin is too thick to be driven off; the lock is round so it's hard to grab, and the lock cylinder is almost impossible to pull for through-the-lock entry.

There are only three ways to defeat this lock: Cut it with a rotary power saw, cut it off with a torch, and apply force with a pipe wrench and snap it off.

The idea for this tool comes directly from the City of New York Fire Department. FDNY faces a high number of this type of lock and uses the hockey puck lock breaker effectively and safely. As the American Series 2000 gained in popularity as a high-security lock, it spread all over the country. The lock is readily available in all communities, and its use is increasing.

There are two basic types of hockey puck lock breakers available to the fire service. The first can be purchased locally at the nearest tool supply or hardware store. The second type is available from manufacturers of fire service tools. Both function equally well, but the tool designed by fire service manufacturers is probably the better bet if you are going to have to use this lock breaker a lot.

The locally purchased item is available in various sizes. The best size is either the 36-inch pipe wrench or the 48-inch pipe wrench, in steel. Aluminum is available, but steel will provide a higher level of safety for what you are going to do with this tool. These steel wrenches are very heavy—the 48-inch wrench weighs 38 pounds.

The wrench available from fire service tool manufacturers is also a pipe wrench, but it is only a 24-inch pipe wrench that weighs about eight pounds.

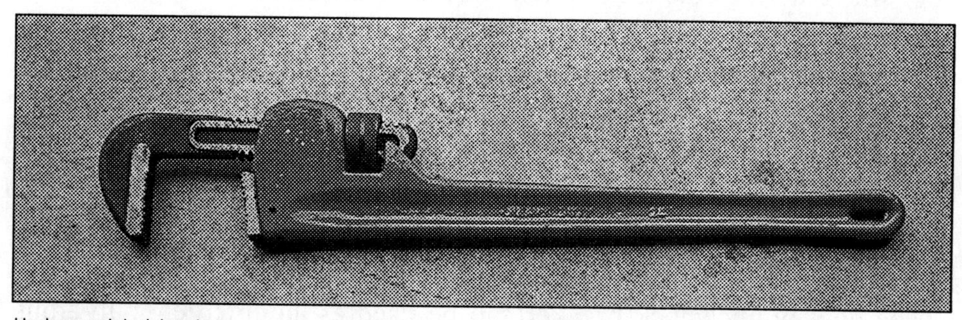

Hockey puck lock breaker.

The fire service model has a bar handle added to it. Yeah, a cheater bar. This is no ordinary cheater bar, however. When manufactured, the bar is tested and built to be "appropriate," or as appropriate as it can be, for use as a handle in a tool designed only to break the hockey puck lock. There may be a million ways to use this wrench, but it was designed for use in forcing the hockey puck lock! Remember that! The cheater bar handle has been bolted and epoxied in place. It is not removable; there are no welds.

A four-foot section of two-inch-inside-diameter galvanized steel pipe can be purchased along with a standard wrench to add a cheater handle for use on high-security locks. If making your own cheater, ensure that the pipe is thick-walled steel and that it fits all the way onto the pipe wrench handle, stopping just under the head of the wrench.

The use of the hockey puck lock breaker is relatively simple, but the most effective use again depends on correctly sizing up the situation. In order to use this tool, the hockey puck lock must be attached to a high-security staple or hasp assembly. If it is attached to a weak or poor-steel type of staple or hasp, the lock will only damage the staple as you try to remove it. Size up the attachment point as well as the lock itself.

Open the jaws of the pipe wrench until they slip over the top of the lock easily. Set the tool in place with the handle at an angle that you can comfortably reach, and have the lock as far back into the jaws as possible. Tighten the jaws on the lock—make them tight. Push the handle of the tool down slightly to ensure that the jaws are going to lock in and bite into the surface of the lock casing. The lock is case-hardened steel and very slippery. If it is wet or ice-covered, double-check the bite of the tool.

When you are sure the tool is set, face the handle of the wrench. Apply downward pressure on the handle in a quick snapping motion. The idea is to snap the lock off the staple. Two firefighters may be needed. Watch the placement of your feet. Do not allow your feet to get under the handle of the wrench. You will be applying tremendous force, and if the tool slips it could crush your toes, right through the steel toe of your boot. The snap motion is important. If you apply slow, continuous pressure to the tool, it will allow the lock to bend and jam. A jammed hockey puck will have to be cut off.

The hockey puck lock breaker is a long lever. The force at the head of the tool where it grabs the lock is tremendous. If the second firefighter is not needed to apply pressure to the handle, that firefighter should stand clear and observe whether any progress is being made by the firefighter working the wrench. If possible, a warning should be shouted as the lock breaks so that the firefighter on the handle can maintain balance and not fall.

Special Uses

The hockey puck lock breaker can be used as an emergency hydrant wrench. Although not highly recommended by the hydrant folks, it will open those tough little guys that haven't been opened in awhile.

In-House Modifications

A hockey puck lock breaker or standard pipe wrench should not be modified in any way. A complete tool can be built in your own shop, but you may be asking for trouble. The cheater bar can be dangerous if improperly used. Check with several tool manufacturers for prices. The cost of a manufactured hockey puck lock breaker is very minimal. If you are going to produce one yourself, don't skimp. Buy the best wrench that is available, and ensure that the galvanized pipe you will be using as a cheater is of high-quality, thick-walled, galvanized steel. Adding friction tape to the handle of the wrench will improve your grip. When adding the tape, don't add so much that your cheater bar won't fit anymore.

Limitations

Before ordering this tool, you should make a few phone calls or, better yet, visit your local hardware stores and area locksmiths. Ask them about the American Series 2000 lock specifically and other common high-security locks. You may find out that there has never been a single one of them sold in your community or, worse, find out that there have been several hundred sold. At any rate, investigate the need for this item before adding it to your inventory. It is a very useful tool but also a very limited-use tool.

Other limitations include:

- It is heavy.
- It will take up a lot of compartment space!
- The jaws and threaded mechanisms must be maintained often to ensure it remains rust-free and easy to move and that the grip surface is sharp enough to bite into the lock body. The tool has a tendency to slip easily.
- Because of its design, this wrench will lend itself to being misused for other functions that may damage equipment or injure firefighters.

A TOOL

Standard Uses

The A tool isn't really a separate tool. An A tool can be machined into a variety of tools including halligan bars, Chicago patrol bars, mini-halligans, officer's tools—even into metal stock purchased at a local supply house.

The A tool is the beveled, triangular-shaped lock puller found in many fire

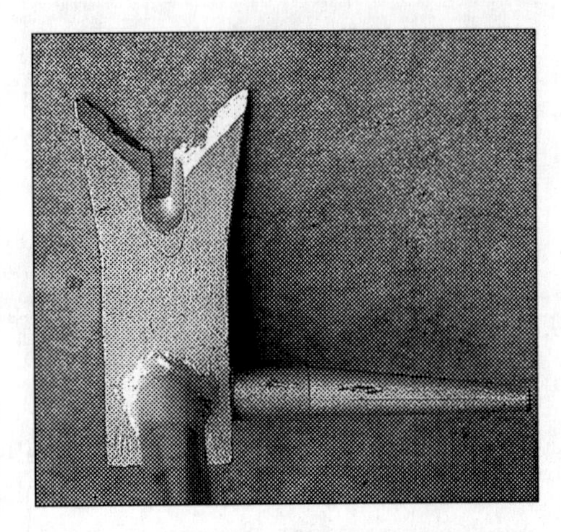
Close-up of an A tool in a mini-halligan bar.

service tools. The original A tool started out as a modified nail puller purchased at a local hardware store. It was developed after lock manufacturers developed methods of preventing the use of the K tool to pull lock mechanisms.

Some A tools perform better than others. A lot depends on the way the tool was machined, how sharp the bevels are, and how proficient the user is. There are locks being developed today to thwart the use of the A tool. It is a good device but not foolproof in every situation. All A tools, no matter what tools they are machined into, work basically the same.

First, you must size up the lock and door. When you determine that the A tool is to be used for through-the-lock entry, you'll need the A tool, a striking tool, and a key tool.

Set the wide opening of the A tool over the top of the lock cylinder to be pulled, and cant the tool at a 45-degree angle. Strike the top of the A tool with a striking tool, driving it down behind the lock faceplate and onto the cylinder itself. You will chew up the door a bit if it's a wooden door. Keep going. You need to set the tool firmly into the lock cylinder body. Most lock cylinders are brass, so the bevels of the A tool will cut into and hold the lock cylinder.

Once you are sure that the tool is well set, pry upward. Always pry upward. If the tool slips before the lock is all the way out, repeat the process. Keep at it

The A tool is positioned over the lock cylinder.

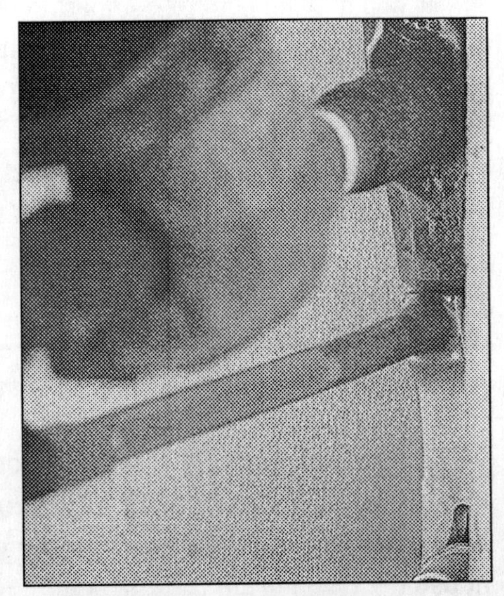
Use a striking tool to tap the A tool firmly down over the lock cylinder. Pry up to remove the lock.

until the lock cylinder is clear of the door. Once the cylinder is out, the lock mechanism will be exposed for you to trip with the key tool.

It is that simple. You'll run into locks that will give you fits trying to pull them out. They may be held in place with case-hardened or oversize screws; any number of things can happen. Keep at it. Even if you mangle the lock cylinder getting it out, as long as the integrity of the lock mechanism is intact, you'll be able to open the door.

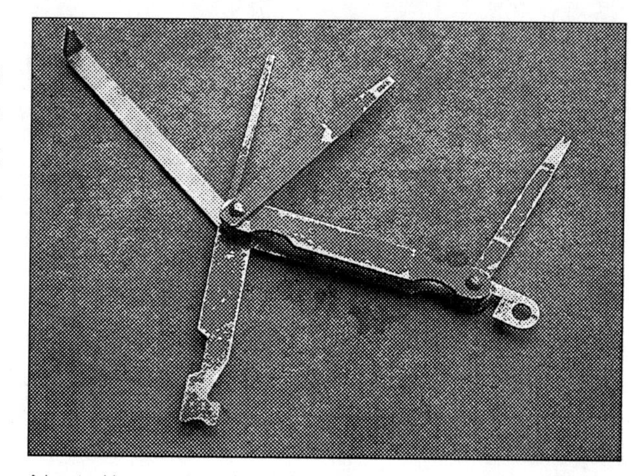

A key tool is necessary when using the A tool for through-the-lock entry.

Special Uses

The A tool is a special-use tool. To keep it sharp and capable of pulling lock cylinders, don't use it for anything else.

In-House Modifications

Don't modify the A tool. You may modify the tool that it is machined into, but don't modify the A tool itself.

Limitations

The A tool is for pulling lock cylinders only. If you use the tool for anything else, it may not pull locks when you need it to!

J TOOL

Standard Uses

This is really a unique tool. The J tool has extremely limited use. There is nothing else you can do with it other than open doors equipped with panic hardware.

The tool was developed by New Jersey firefighters. It is quarter-inch stainless steel bent into a J shape, with an angle brace at the bottom corner of the J. Its only purpose is to slip between the weather stripping of commercial doors and to trip the panic bar. Simple. It works best on doors equipped with panic bars, but it will also work on paddle equipment. If you get good with it, you can

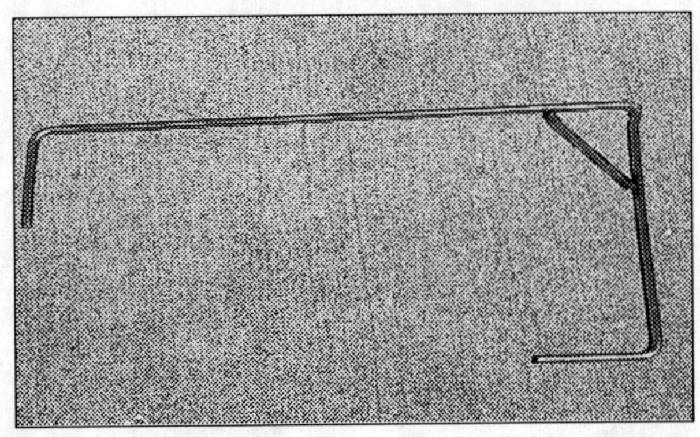
J tool.

even trip those panic bars that are slightly recessed into the door.

Size up the door. If it is a double door with the opening at the center, even better. Insert the tool, J-side up, a few inches above where you estimate the panic bar to be. Push the tool in deeply. (Hey, if it's a glass door, look through the glass and watch what you're doing. If it's a panel door, read on.) Rotate the tool 180 degrees so that the J side is now down. Slowly and firmly pull the tool out toward you. The J will catch the panic hardware and release the door—that is, if the door hasn't been chained shut or has a door-club device installed to prevent you from using the tool you have in your hand.

Great tool. It's easy to make at the firehouse and can really come in handy at those smells-and-bells calls when there is no lock box and the key holder is 20 minutes or more away. You're in, you've investigated, you're history!

Special Uses

There are no special uses for the J tool.

In-House Modifications

The best in-house modification that you can do with a J tool is to make your own. All that is needed is quarter-inch round stock that you'll bend into a sufficient-enough J shape to be effective. Make a couple in some different configurations. Try them out and keep the one that works best.

Limitations

Sometimes the tool will work, sometimes it won't. If the door has been chained shut, or if another security device is used, the tool is useless.

K TOOL

Standard Uses

The K tool was the first tool available to firefighters that was commercially machined to pull lock cylinders for through-the-lock forcible entry. The K tool has a steel body with a metal strap across its back to accept the adze or other surface of a prying tool. On the face of the tool are two replaceable beveled blades that form the letter K. One side of the K is a little bigger than the other, allowing for different sizes of lock faces.

K tool.

The K tool is a very limited-use tool today. The tool was invented and patented by a lieutenant in FDNY who was also a licensed locksmith. When it was first developed, the K tool was able to pull most lock cylinders. Well, burglars discovered it, and lock companies immediately developed locks to defeat it. Collars and other devices were installed on locks so that they would not fit into the K tool. It still is a very viable tool, though, and will pull a great number of locks currently in use.

As with any other forcible entry situation, size up the door you are going to try to enter. Determine which way it swings, what material it's made of, and what type of device is holding it closed. If it has a lock that has a very narrow profile in the door, the K tool can be used. Many doors in commercial establishments such as strip malls and shopping centers still use this type of lock.

To use this tool effectively, you need the K tool, a key tool, a striking tool, and a prying tool. The prying tool should have an adze on it. Get all the tools you need at the door. Grab the K tool and slide it over the face of the lock so that the lock is between the two blades. Put it on whichever way it will fit. It can be perpendicular or horizontal to the door. As long as the lock is in between two of the cutting blades of the K tool, it doesn't matter which way the tool is aligned.

Set the tool with your hand as best you can. Now, gently tap the top of the K tool to drive it down on the lock. Use the back of a flathead axe or other striking tool. The blades need to cut deeply into the brass body of the lock because the blades are actually going to do the pulling. All the force you will apply with the prying tool is going to be transferred to the K tool's blades. The tool is going to chew up the lock a little, so make sure the tool is well set by looking into the open end of the tool to see if the blades have cut into the lock cylinder. If they have, you're ready to pull the lock out.

Insert the adze of your prying tool into the strap on the back of the K tool. Insert the prying tool in a way that the K tool isn't going to fall off and hit you in the foot when the lock pulls free. Usually you insert the adze of the prying tool from the bottom of the strap, but it depends on which way you drove the K tool onto the lock. Don't let it fall off and hit you. The K tool is heavy, and it hurts when it hits you.

When you are set, get a firm, feet-apart stance. Move to the end of the pry bar and pry upward. Prying downward may damage the lock mechanism inside the door. If you damage the lock mechanism itself, you won't be able to open it with the key tool.

The lock should pull out when you pry. If it pulls past the K tool and is still hanging mangled in the hole in the door, don't worry; just repeat the process. Some locks are tough, so don't get frustrated—do it again. Repeat the process until the lock is out. When the lock clears the door, use the key tool to trip the mechanism.

Special Uses

There have been occasions when the K tool has been used to shut off gas utilities. Insert the gas valve into the K tool. Either use a halligan bar or other pry bar to turn the valve to the *off* position.

In-House Modifications

Keeping the K tool and the needed key tools together is a problem. One solution is to add a chain to the K tool and attach the key tools to the chain.

Limitations

- It won't work on mortise or rim locks with collars.
- It is carried in a separate pouch as a separate tool, so you don't have it with you in hand.
- It's easy to lose or misplace.
- It requires a lot of practice to be efficient with this tool.
- Once the cutting blades are damaged, the tool won't work.

DUCK-BILLED LOCK BREAKER

Standard Uses

A what? A duck-billed lock breaker! The DBLB is a neat tool although extremely limited in use. For firefighters it is an effective tool, especially in today's security-conscious world. The DBLB is a wedge of steel on a handle for breaking open the shackles on padlocks. It's a very simple tool, available commercially; otherwise, you can make one at the firehouse. If the shackle or bow of a padlock is visible, you'll be able to force the lock open using the DBLB. It is an extremely simple operation.

Size up the padlock. If you cannot attack the device the lock is attached to, use the DBLB and attack the lock. You'll need the DBLB and a striking tool. The striking tool should be fairly heavy, at least eight pounds.

Insert the narrow point of the DBLB into the shackle of the lock. Push the tool down firmly. Make sure your hands are away from the head. Strike the backside of the DBLB with the striking tool—smack it hard! You are driving the wedge of the DBLB deeply into the space between the shackle and the lock body. Keep driving. Make sure your feet are not underneath the area where the lock will fall.

The lock shackle will fail. The shackle will break and pull out of the lock body, and you're in. It's an incredibly simple operation. The DBLB works on almost all locks that have a shackle or bow. Some locks will require more strikes to get them to give than others, but you will cause them to fail.

Duck-billed lock breaker.

If narrow enough, the DBLB's bill can be inserted into car trunk locks to drive them out.

Special Uses

There are several special uses for the DBLB. It can be used to break windows covered by wire mesh. These windows are typically found in factories or in schools and other public buildings. The point of the DBLB is inserted into the mesh, then the tool is driven with a striking tool. As the tool is driven in, the mesh will widen. When the point comes in contact with the window, no more glass.

If maintained well, the tool can be used to break thermal pane windows. Drive it with a striking tool to limit the chances of your hand going all the way through the window. Driving the tool in will break all the layers of the glass.

To some degree, and only if properly constructed and maintained, the DBLB can be used as a striking tool.

In-House Modifications

There are some modifications that must be made to the commercially available DBLB. First, the tip of the tool must be narrowed down to fit some padlocks. The commercial tool is much too wide at the point to fit into standard padlocks. I've even found some high-security padlocks that it won't fit into.

Use a file to take down all the sides of the tool, from the point up and back toward the handle for at least two inches. File the tool so that it fits into a standard, run-of-the-mill lock you buy at the hardware store.

You can make your own tool easily. I've seen them made of steel or brass. The material must be really tough stuff. You'll be trying to break some pretty sophisticated locks made of high-grade material. If you build a chintzy tool, you'll get chintzy results.

My DBLB is made out of one-inch steel that was cut from an old steel door that was in a fire training burn building. A four- by 10-inch rectangle was cut from the door. We measured over about three inches on the short side and, using a torch, cut an angle down to the opposite bottom corner of the steel block. Once the steel had been cut into a triangle and cooled, I cleaned up the edges with a grinder and file until I had very square edges all the way around the tool. Then I got a lock and guestimated how wide the tip if the tool would have to be to fit. Again using the grinder, I whittled the tip of the tool down until at least two inches of it fit into the space between the shackles of the lock.

We then welded a steel 18-inch handle onto the bottom, flat side of the tool head. I cleaned that up with a grinder and file. I sanded the whole thing down, then primed and painted it. When the paint was dry, I added French hitching to the handle and added the tool to my inventory. I've used it several times, and it works great.

You may want to add some length to the handle if you're making your own. A handle of up to 30 inches long would allow the firefighter holding the tool to be out of the way of the firefighter striking with an eight-pound axe or sledgehammer.

To make the tool more versatile, add a pipe wrench to the handle. The tool can be flipped over and the DBLB can be used as a hockey puck lock breaker for round locks or even standard padlocks that are too small to have the DBLB blade inserted.

Limitations

- Its use is limited to breaking locks with shackles.
- It is heavy.
- It is easy to lose once gaining entry.
- To be safe and effective, it's a two-firefighter operation.
- The same operation can be accomplished with a halligan bar.

RABBIT TOOL

Standard Uses

Warning: An excellent point of safety was made by the members of Ladder

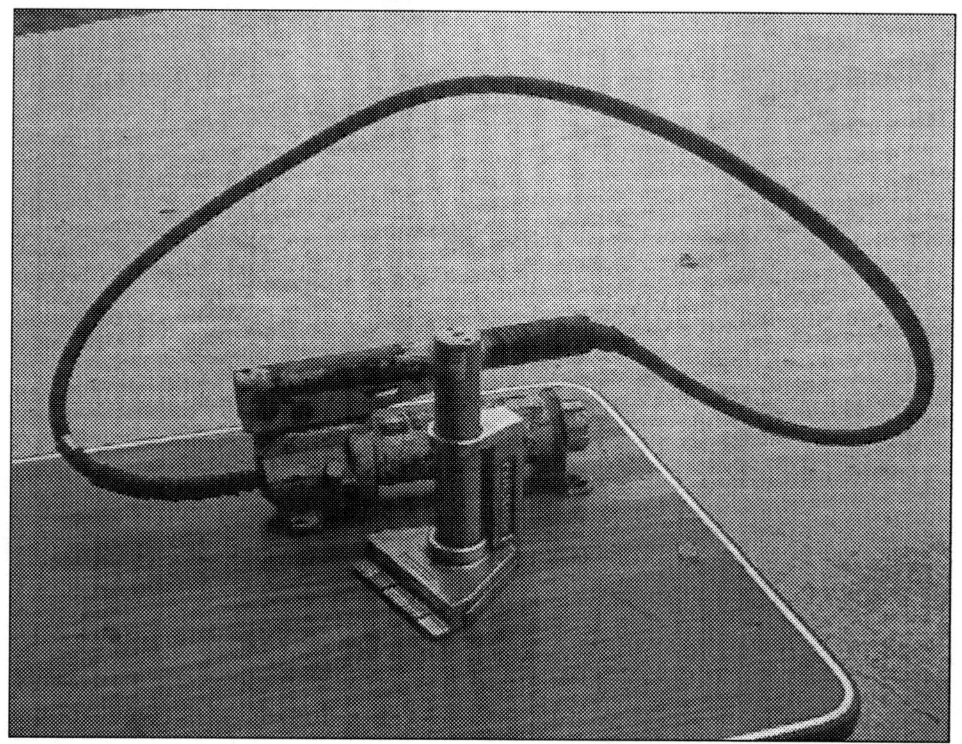

Rabbit tool.

Company 56 and Engine Company 48, FDNY. The rabbit tool will allow entry into fire buildings and rooms with amazing speed. Doors will be forced and truckies will be in and searching, or engine companies will push in and attack the fire before the firefighter assigned to the critical job of vertical ventilation has even reached his operating position. Companies will be operating in superheated, unventilated areas because they were able to gain access so quickly. It has been suggested that this may be one reason that so many firefighters are being caught in flashovers and suffering terrible burn injuries. Exercise caution when using the rabbit tool or any tool or tool combination that will allow rapid entry of firefighters into unventilated areas.

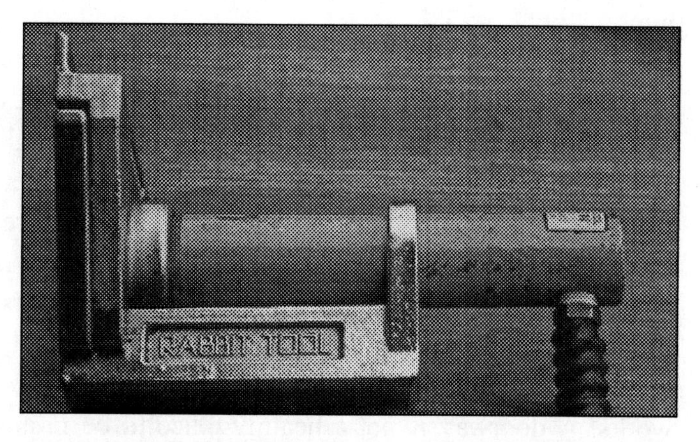

The rabbit tool is a name we use in the fire service to cover a broad range of hydraulic tools used to force open doors. The rabbit tool is the best forcible entry tool that any fire department can own. Technically, it probably doesn't belong in this book. This is a hydraulic tool, pumped by hand. It is similar to the portable hydraulic tool so common in auto extrication.

I've included it here because I classify it as a special-use hand tool. There are no motors or electric pumps or power sources used other than the handle to pump the hydraulic cylinder. This tool is the nuts.

There are many different types of rabbit tools out on the market today. They have a zillion different brand names, and they all have different claims. The one that we're looking at is the basic rabbit tool operated by two firefighters.

The tool consists of the spreader jaws and the pump. They are connected by a high-pressure hose at least four feet long. The spreader jaws have a wide, flat top capable of spreading tremendous force in all directions from the head. A set of teeth or a beveled edge is on the side of the tool, to be inserted between the door and the doorjamb.

The two-firefighter tool is important. Manufacturers are producing tools to be used by one firefighter. They may look great, have lower manpower requirements, offer ease of operation, and all the rest, but they also put the single firefighter at greater risk when opening a door. To open a door with a one-firefighter tool, you have to be standing or kneeling in front of the doorway to set and operate the tool. The tools are extremely effective, and when you pump the tool, the door opens. If there is fire on the other side of the door or a backdraft potential, the firefighter who operated the tool is toast. An additional problem occurs when forcing doors that have locks equipped with shield plates. The bolts from these plates become bullets when forced.

The two-firefighter tool allows the tool head to be set and both firefighters to get out and away from the door before it opens. One firefighter can control the door with a rope or vice grip and chain as the other firefighter pumps the handle. Another possibility is for one firefighter to pump while the other firefighter reaches out and holds the tool to prevent it from falling, but he should also use the wall as a shield in case something tries to jump out and eat him.

In some areas—and this problem is not limited to urban areas—the rabbit tool will be used to force doors in apartment complexes, motels, and so on where drug dealers may be present. Whether a firefighter bangs on the door and announces his presence is not enough. A firefighter who gets no response from within a room that must be searched will break down the door to accomplish his mission. On the other side of that door may be a huge stash of illegal drugs and armed crazy people. When the door pops open and there the firefighter stands holding the one-firefighter rabbit tool, he may not have time to duck the spray of gunfire from within. The rabbit tool makes no noise as it works! A doorway is not a healthy place to be in any emergency situation except maybe an earthquake.

The two-firefighter operation is very simple. Size up the door. The rabbit tool only works on inward-swinging doors. Insert the jaws of the tool between the door and door frame right on top of the lock. Lean on the door a little to get the jaws of the tool set. If you have to, use a striking tool to tap the rabbit tool in place. It won't take much. Once the tool is set, do something to control the swing of the door. Attach a utility rope or a vise grip and chain to the door so that you can pull it shut again if fire jumps out at you and threatens the hallway.

Get back and away from the door. One firefighter can reach out and hold the tool to prevent it from falling to the floor when the door opens. You can grab it near the hydraulic hose to keep your hand out of the danger area. The second firefighter makes sure that the release valve for the hydraulic fluid is in the closed position. Start pumping the handle of the hydraulic pump. Four pumps is usually what it takes to open the door.

This is a very simple and effective operation. If you have any hotels, motels, apartment complexes, or multiple-dwelling structures, a rabbit tool should be in your inventory. They're expensive, but they are really incredible tools when multiple doors must be opened.

Special Uses

The rabbit tool has a special use in auto extrication. The tool operates very smoothly and very quietly. The jaws of the rabbit tool head can be used to spread open doors, hoods, and trunks on vehicles to give the bigger hydraulic tools a purchase point. Using the rabbit tool prevents the hydraulic tool operator from having to slam the spreader tips of the tool he's using to get a purchase point to begin extrication. It is much easier on the patient inside the vehicle.

The rabbit tool can also be used to open hoods and trunks on vehicles for fighting fires. We've all messed with those pesky hood and trunk latches trying to beat our brains out getting them open. The rabbit tool will spread the sheet metal far enough to get a hoseline into the space to extinguish the fire. Then we can get the tools or widgets or whatever we need to trip the cable to open the hood or trunk normally. At least the fire will be out.

The rabbit tool is the first tool of choice for opening elevator doors. The halligan bar is second.

In-House Modifications

Do not modify the rabbit tool. There are, however, several other tools that may be added to the kit to make you more efficient. FDNY Ladder 56 in the Bronx, New York, has added a 25-foot length of rope to the rabbit tool kit to use to control the door. If fire jumps out, the officer can pull on the rope to close it quickly. Additional rubber latch straps have been added to the kit to keep forced doors from relatching should they close while firefighters are inside searching or after they have moved down the hall and the engine company needs to move in.

Limitations

- The rabbit tool is expensive.
- It only works on inward-swinging doors.

ROOF CUTTER

Standard Uses

The roof cutter is a tool that has been around for so many years nobody can remember where it came from. This tool comes in a variety of styles and models, but it is basically a can opener on a handle.

There are models with rollers; models with chromed, crescent-shaped cutters; and other styles. They all function the same way: They are designed to cut sheet metal the same way a can opener is designed to cut open the top of a metal can. These tools are found in many fire departments. The roof cutter is one of those tools that you just can't throw away—even though you never use it, it might come in handy sometime. Most tools are forged steel cutters on a 40-inch D-handled hickory wood handle. When maintained properly, they are incredibly sharp and, believe it or not, very effective, although manpower-intensive.

Roof cutter.

To use the tool, size up the situation. What are you cutting into, how big does the hole have to be, and what obstacles are you going to run into? The tool is very effective on mobile homes, rural outbuildings, tin roofs, and other light metal. In certain cases, you may be able to peel back the skin of an automobile.

Slam the tool down into the material you're going to cut. Try to insert the blade along a joist or stud, and use that for support. The cutting head is shaped to give you a fulcrum. Push the blade straight into the material, sinking it all the way to the curved fulcrum point. Exert forward pressure on the handle and pull upward. As you rock the blade forward, it will cut the metal. Jam the tool forward again and repeat the process until you have the hole open. Be very careful where you step. Don't step in the cut area or you'll fall in!

Tools with rollers work a little bit more easily than the can-opener type. With the roller type, set the blade of the metal cutter into the material you're cutting. Push forward on the handle of the tool, and wheel the cutter along.

It's supposed to cut and peel back the metal as you go. As before, watch your step.

While this is an obscure tool, you should still practice with it. In an extreme emergency, it may be the only tool you have to effect a rescue. In extreme emergencies, you may be without any power sources. (Recently there have been entire fire departments wiped out by tornadoes. Fire station, apparatus, tools—all gone.) This tool is like an axe. It always starts. It doesn't need gas or electricity. It may be old, but learn how to use it. You may be called on to use it at the next natural disaster when there is no power, no gas, and every possible temporary power source is being used at city hall. Think about it.

Special Uses

There are no special uses for this tool.

In-House Modifications

If you have one of these tools, you may want to consider updating it a little. If it still has its original hickory handle, inspect it for defects. Change it for a sturdier fiberglass one with a D handle. To increase the leverage of the tool and save your back, lengthen the handle to about 45 inches and possibly to 50 inches. Maintain a very sharp cutting edge.

Limitations

- It's a slow method of opening up metal.
- Firefighters may never have used it or obtained enough training to use the tool effectively.
- It is a very sharp and dangerous tool to carry incorrectly.
- It should be the first addition to your fire museum. There are many more efficient tools to choose from.

SHOVE KNIVES

Standard Uses

Every firefighter should have two shove knives in his pocket. The shove knife is an excellent tool for springing the latch on outward-swinging doors or the spring latches on double-hung windows. It is a must-have tool if you are working in buildings that have self-locking doors in smoke stairs or other egress routes. The shove knife will let you back in should the door close behind you. Shove knifes can be made in the firehouse or purchased commercially. Either way, they are cheap and handy.

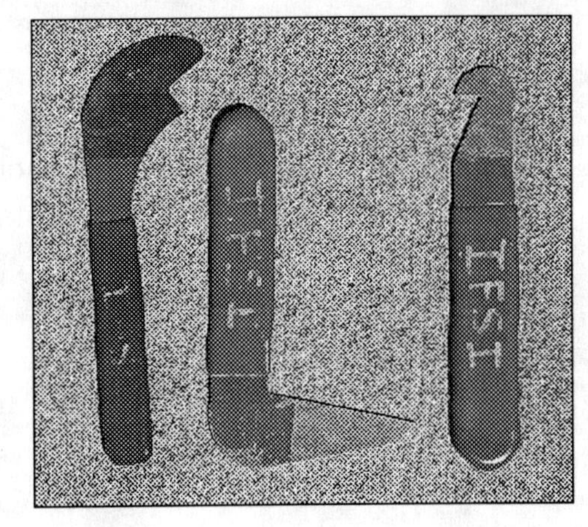

Shove knives.

To use a shove knife, size up the door you need to open. If it is locked with a dead bolt or other secure lock, you can't open the door. I don't care that they do it on television with a credit card—it won't work, so find another door.

If the door swings outward (toward you) and is latched with a spring latch, insert the shove knife between the door and jamb, above the latch. Slide the tool downward, and engage the spring latch with the notch in the shove knife. Slide the shove knife down until it releases the spring latch from its keeper, then pull the door toward you. You're in.

If the door is an inward-swinging type, insert the shove knife behind the stop molding, above the spring latch. Slide the tool downward and engage the spring latch with the notch in the shove knife. Slide the shove knife down until it releases the spring latch from its keeper. Push the door open.

Special Uses

You can also use it in other forcible entry situations for sliding underneath a door to see whether a multilock or a door club is locking the door.

In-House Modifications

The shove knife can be made at the firehouse using thin spring steel, banding material, or other thin flexible metal. I've even seen some cut from galvanized metal. The notch is similar to the notch in the old slim jim car door openers.

Limitations

- If the door is secured with any lock or latch other than a spring latch, the shove knife won't work.
- It is best to have two shove knives with you; two work more efficiently than one.

VISE GRIPS AND CHAIN

Standard Uses

A set of vise grips with either a chain or dog leash attached is a tool that has a

Vise grips and chain.

multitude of uses on the fireground. Until you make a set for yourself, you really don't know how many times it would have been nice to have had a set available. The tool can be made any way you want. I recommend that you have a small personal set for yourself in your pocket and a larger, heavy-duty set or several sets on your apparatus.

To make this vise grip and chain set, go buy a good set of vise grips. Not cheap ones, good ones. Your life may be depending on this tool, so don't be cheap. Next, stop at the local pet store and buy a chain dog leash suitable for a small dog. Drill a hole in the top handle of the vise grip, sufficient to allow the snap hook of the dog leash to fit and swivel around a little. It's done. Wrap the leash around the vise grip, and slip it in your pocket.

A heavier-duty set should be made also. Get another vise grip. This time while you're in the hardware store, buy about 24 inches of dog chain, the kind you attach to a black lab. Also get a U bolt. Put the first chain link on the U bolt. Weld the U bolt to the top jaw of the vise grip, just behind the curve of the head. A straight steel bolt with the head cut off and ground smooth can be welded through the last link of the chain, making a T handle. Once you have the tools made, you are ready to put them to work.

The vise grip and chain can be used in conjunction with all of the forcible entry tools in this book. When performing forcible entry, it is critical that you be in control of the door at all times. Inward-swinging doors are especially important. By attaching the vise grip to the doorknob and hanging on to the chain handle, you are always in control of the door. When the irons man pops that door open, you will be able to pull it shut again if something nasty jumps out at you. Getting the door open is important, but keeping the fire out of the hallway and away from you is even more so.

When using a power saw to cut padlocks, the vise grip should be attached to the padlock and the lock pulled out straight and tight so that the saw operator can cut both sides of the shackles at the same time. Being able to cut both sides is critical. High-security locks will not open unless both sides are cut. On vise grip and chain setups for saws, consider a longer chain. I have a 36-inch chain on mine because I deal with students learning how to use the saw. I stay well clear of the inexperienced firefighter holding on to a saw with a metal cutting blade spinning at 6,000 rpm!

Special Uses

There are no special uses for the vise grip and chain. It is a limited-use tool.

Limitations

- They're bulky to carry around in your pocket.
- You don't always have enough to go around at large operations.
- Vise grips get misplaced. Usually they're found in the toolbox in the boiler room, placed there by some unsuspecting person who didn't know what they were for.

BATTERING RAM

Standard Uses

If ever there was a truly ancient tool currently seeing a revitalization in the fire service, it would have to be the battering ram. This tool dates back to biblical times. It is a weapon of war, a tool for mass destruction. It was designed to batter down the walls of cities so troops could pour through the

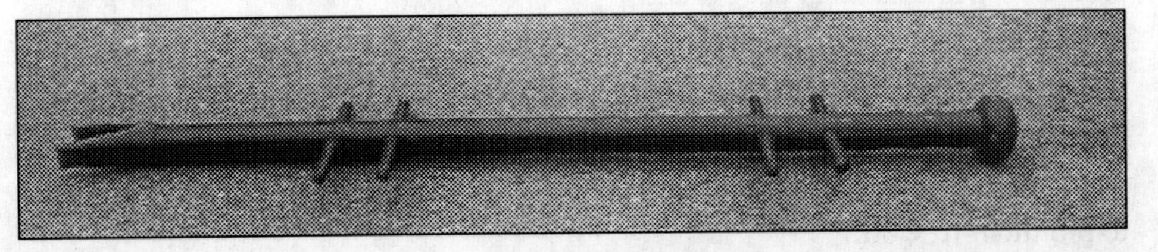

Battering ram.

breach to loot and pillage the citizens. We still use it in exactly the same way today. Well, all but the last part.

The battering ram is a heavy steel pole with a rounded head. It has handles along the side to be used by two to four firefighters for bashing in masonry walls to make a breach to access the fire. A more modern version of the battering ram is probably kept at your local police department. Law enforcement S.W.A.T. teams have taken the ancient art of battering down walls and doors and have perfected it. The local police department may have a bunch of one- and two-officer battering rams to use during drug raids.

Using a battering ram takes practice and teamwork. It isn't enough for one firefighter to know how the tool functions and what the goal is—everybody has to know. Using the tool is relatively simple. Swing the ram back and forth like a pendulum, striking the masonry surface, knocking it down. The use of the battering ram calls for expert size-up, plus a thorough knowledge of building construction and Newton's Laws.

The battering ram is an excellent tool for opening holes in masonry walls. Sledgehammers should be used to finish off and enlarge the hole. The battering ram is very unforgiving and good for the rough work.

Two firefighters can swing the tool, striking the wall at the juncture of at least four bricks. The momentum and force will easily knock out the bricks. There is an art to using it to open holes in a wall. You must be very conscious of building construction and what is holding the building together. Freelancers using the battering ram wherever it is convenient, rather than considering building construction, are asking for a catastrophe. Firefighters using the ram should work together and build a rhythm for swinging this heavy tool.

Battering rams can also be used to force down steel doors. If the door is heavily secured and there is no possible way to gain entry using conventional or through-the-lock forcible entry, then a battering ram may be the tool of choice.

The main problem with using a fire department battering ram for opening doors is space. There isn't a lot of room in the vicinity of most doorways, so firefighters are going to be crowded. Onlooking firefighters need to be warned against standing directly behind the swinging tool. If it bounces or slips from the operators hands .... Well, the tool is very unforgiving.

Another problem in using the tool on doors is that the head of it may plow a hole through the door rather than batter the whole thing down. Police departments have added protective collars to their rams to minimize the chances of going through the door.

Just as with any forcible entry tool, the idea is to concentrate on what is holding the door closed. The battering ram should strike the door as close to either the lock side or the hinge side as possible. Swing the tool like a pendulum, allowing its mass and momentum to do the work. You will not have a door left when it finally gives, so be prepared for what's behind it!

The battering ram is the most ancient tool we have in our toolbox. It isn't often used, but it shouldn't be forgotten when a need for extreme force arises on the fireground and there is no power equipment suitable for the job. Many truck companies across the United States still carry a battering ram; it is still inventoried each day and painted each year during spring housecleaning. It is a special-use tool, but you have to know how to use it to be effective.

Special Uses

The battering ram has no real special use in the fire service other than to knock holes in walls, open stubborn doors, and accomplish other heavy-force tasks. The local law enforcement agency in your jurisdiction may have some real special uses for it. Additionally, they may have a wide selection of smaller, more manageable rams that could replace the behemoth you now have on the rig.

In-House Modifications

There isn't much you can do to improve on a several-thousand-year-old design that probably hasn't already been tried and abandoned. A removable protective collar for use when battering down doors would be an advantage. A permanent welded collar would be mashed flat the first time you drove the tool through a masonry wall. The collar would help prevent the tool from punching a hole in the door and distribute the force more equally in all directions to make the tool more effective.

Limitations

- Modern hydraulic or pneumatic tools have almost made the battering ram obsolete—not quite yet, but almost.
- It is very heavy and requires a minimum of two firefighters.
- It requires training and practice to use.
- Its use must be supervised by a firefighter or officer thoroughly knowledgeable in building construction.
- Whatever you hit with it is destroyed.
- It isn't usually stored on the apparatus in a convenient location.

COMBINATION PUNCH AND CHISEL

Standard Uses

This is a unique and very obscure tool that was carried by many fire departments for years, and then it disappeared. It disappeared because it had been replaced by more useful hand tools, such as the halligan bar. The combination punch and chisel is an elongated S-shaped brace that holds a heavy cold chisel at one end and a heavy steel punch at the other. The brace allows a firefighter to hold the tool out and away from the other firefighter swinging a sledgehammer.

Many East Coast departments carry this tool. It is not often used. It is a handy tool for dealing with very heavy materials often found in mill-type construction, piers, large factories, rail yards, and shipyards.

The punch is used like a normal punch, only on a larger scale. It can be used to drive rivets out of steel, start holes for saws, and chip mortar and cement. The chisel can be used to shear bolt heads, rivets, and screws; to split heavy planking; to start tearing up flooring; and to knock heavy steel hinge pins out of doors.

This isn't a common tool anymore. Nevertheless, firefighters still run into those old, full-dimensional lumber places that the common pry bars, hooks, and other tools just can't get started. Having a combination punch and chisel can be an advantage because they work well to split and start the overhaul process when you just can't get a purchase any other way.

Special Uses

The standard use of this tool is its special use. The tool is used when standard firefighting forcible entry or overhaul tools are ineffective in obtaining purchase points or in removing metal structures, rivets, and so on.

Limitations

Today's firefighting hand tools provide most of the capability of the punch-and-chisel combination.

HAMMERHEADED PICK

Standard Uses

The hammerheaded pick is a very old tool that came into the fire service via the railroad. The tool is very simple: a striking or hammerhead surface on one side of the tool head and a pick on the other side. The pick is wide at the head and tapers to a point at the end.

The tool was originally used by railroaders in laying and maintaining track. It is very difficult to find today, since the era of maintaining track by hand is long past. I have, however, found several of these tools in fire departments.

The primary use for it in the fire service is as a digging tool. The striking surface as well as the pick can be used for digging earth, as in trench rescues, or for dismantling concrete or block in masonry accidents, such as collapses.

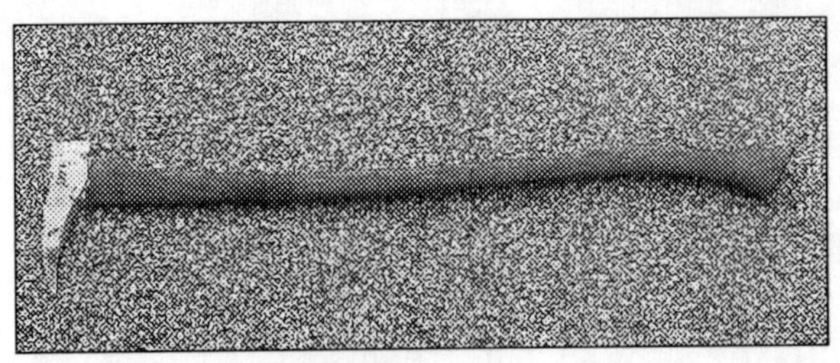

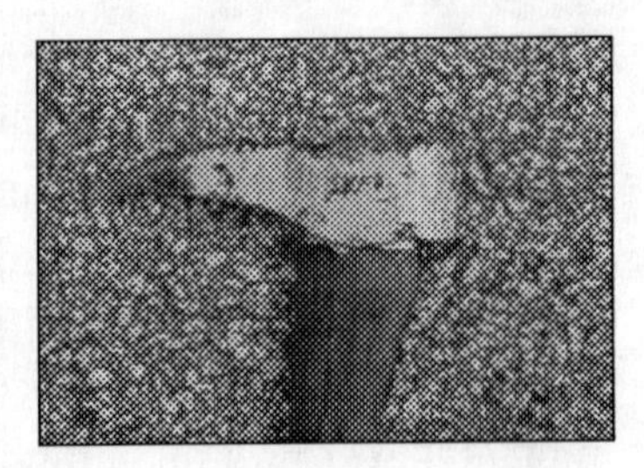

Hammerheaded pick.

This tool is very efficient, but in most cases it is overlooked because it is not new. The hammerheaded pick usually has a solid ash handle, and when properly applied, it will make quick work of breaching concrete or busting up rubble.

The tool is designed as a striking tool, so it can also be struck. The tool can be set against an object like a big piece of stone or concrete, then struck with a striking tool. The force applied will usually drive in the pick and shatter whatever you're trying to break up.

Special Uses

The neatest use for this venerable old tool is as a lock breaker for modern-day, high-security padlocks. To break locks with this tool, insert the pick of the tool into the lock. Make sure the pick is between the lock shackles. Strike the hammerhead of the tool with another striking tool. Continue to drive the pick through the shackles. The lock will eventually give as you drive the widening blade of the pick inward.

The long handle of the tool allows one firefighter to be back and away from the lock while another firefighter strikes the tool head. It also allows both firefighters to be away from the lock when it fails. The lock usually fails without warning, and it flies downward toward the ground with tremendous velocity. Dismembered locks will break your foot, fire boot or no fire boot!

In-House Modifications

To keep the tool working for many years to come, add some overstrike protection to it, as described on page 10.

Special Uses

The special use of the hammerheaded pick is as a lock breaker, although I am sure that the original engineer of this tool didn't have that intention in mind. It is a very effective lock breaker, as well as an effective tool for use in collapse or trench rescue.

Limitations

- The tool you have is probably old and the handle will be rotten. Make a thorough inspection of it before placing it in service.
- Handle replacements may be difficult to obtain.
- It is a very limited-use tool. It affords no real prying or pulling capabilities.

REBAR WINDOW BREAKER

Standard Uses

This tool was submitted to me by Firefighter Michael N. Ciampo of Ladder Company 44, FDNY, in the South Bronx. This is a simple homemade tool used to break windows through wire mesh screens, often found on factories and school buildings. Ventilation can be accomplished without having to remove the mesh, which can be a very time-consuming process. It also simplifies the process by eliminating the need for firefighters to smash away with sledgehammers, trying to bend the mesh enough to smash the glass behind it.

The tool is 4½ feet of rebar. One end is tapered to a sharp point; the other is heated, then curled around to form a handle. The handle must curl all the way around to form a circle. The circle should be large enough to fit a gloved hand. The handle provides protection for your hand.

This tool is simple to use. Insert the pointed end through the mesh and strike the window. Break the glass. Move the tool around as needed to clear out as much of the glass as possible.

Ladder Company 38 of FDNY, also in the Bronx, uses a bent piece of rebar as a heavy-duty J tool for reaching through steel security gates to release the panic bar and open the gate.

Special Uses

There are no special uses, yet.

In-House Modifications

After making and trying the tool, there isn't too much you can do to improve on the basic design. On the tool I made, I ground a rounded point into the tip to give me more of a concentrated pressure point to break the glass. I'm not sure it was worth it, since the point dulls quickly.

You may want a parade-piece tool. Use round stock rather than rebar to make it. You can paint it and it will look a little prettier, but it won't function any differently.

CHAPTER 9: TOOL MAINTENANCE

The effectiveness of any tool used on the fireground is a direct result of the skill of its user. Conversely, the operator's performance with that tool is influenced by its design and the state of its maintenance. Even the best firefighters turn in mediocre performances when their tools have been poorly maintained.

Many firefighting tools are poorly maintained because of a disease among firefighters known as "synthere." Each morning, the tool compartment doors of apparatus all across the country are opened, and the firefighter doing the morning apparatus check peers in.

"Yep, synthere!" can be heard as the compartment door slams shut. The tool "is in there," but is it ready to go to work?

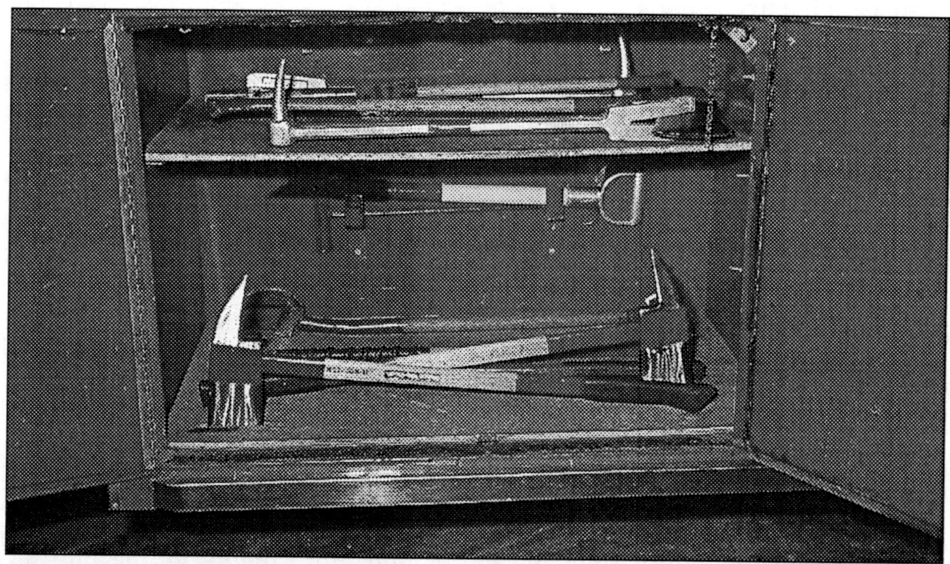

Your tool bin should contain well-maintained tools.

Maintenance must become more of a habit than just making sure that the tool is in its assigned place on the apparatus. Your life and the lives of other firefighters and civilians depend on those tools. Not maintaining a hand tool is the same as not putting fuel in the rig. When you need it to function, it will fail.

In this chapter, we'll look at some of the aspects of tool maintenance. This chapter does not have all the answers about tool maintenance. Included are sound basic techniques for keeping your tools in the best shape and ready for immediate use. When in doubt as to how a tool should be sharpened, cleaned, shaped, or maintained, call the manufacturer. If the manufacturer can't give you satisfactory answers, contact someone who has experience in tool maintenance. Contact local sharpening shops, hardware stores, or even the local library for information on how to maintain hand tools. There are many conflicting ideas on tool maintenance. When in doubt, ask.

CUTTING TOOL HEADS

Fire service cutting tools fall into a gray area. How sharp is sharp? Sharp in the fire service sense is a cutting edge that will not chip or dull quickly and will still cut wood and metal fast and efficiently.

Axes

Remove all the paint from the tool head. Paint hides defects. Polish the tool head using a buffing wheel and rouge or other polishing compounds. The finish doesn't have to be a mirror finish—just clean and free of any scratches or gouges that would allow rust to form.

Sharpen the axe using a file and a hone. Don't use a bench grinder or hand grinder on it unless you are an expert at tool sharpening! A single-cut or double-cut mill bastard file is good. Using slow strokes, file all of the chips, dents, and dings out of the cutting edge of the blade. Follow its original contour. When filing, you should see a buildup of metal shavings forming along the edge of the axe blade. When you see this, turn the axe over and file on the other side of the blade edge. Form another ridge of metal shavings just like the first. The relief of the blade is what will prevent the axe from skipping and jumping around while you are using the axe. The relief, or general bevel of the blade in relation to its thickness, also causes the wood chips to fly out of the way when cutting, allowing you to actually cut material rather than to recut chips.

When you have achieved a nick-free cutting surface, wipe the metal filings off the blade. Now, use the hone to sharpen it. The hone is a finer abrasive material that will remove the file marks and more metal, increasing the sharpness of the blade. Do not make the blade razor sharp! Sharpen it to the point where there is a sufficient relief, with no nicks or dings in the cutting surface.

On pickhead axes, maintain the shape of the pick with the file. Maintain the point by filing along the contour of the pick, removing any deformed metal flakes or shavings that appear on the end. Maintain the edges with sharp, square corners. Remove any chips or nicks.

After you have filed, honed, and polished the axe head, wipe the tool with machine oil or light motor oil. Do not use materials containing 1,1,1,trichloroethane! This product deteriorates wood and resins (glues) of the handle!

Bolt Cutters

Bolt cutters are not self-sharpening tools. Use the file and hone on the bolt cutter as you did on the axe. Unlike the axe, the bolt cutter blade should be honed sharp. The bolt cutter functions initially by biting into the metal to be cut, then shearing through it. File out any nicks or dings, then hone. Follow the original contours and angle of the blade. If the blade has very deep nicks, replace the cutting blades altogether. You can file away too much metal and ruin the blade surface if you are not careful. Light filing and honing are all this tool should need to stay sharp. If the tool has been misused, replace the blades.

Oil the tool with light machine oil or light motor oil. Make sure that you oil the moving joints of the bolt cutter so that it opens and closes with ease. Wipe a thin coat on the cutting blades. Make sure that you don't allow any excess oil to run down the tool and get onto the handles. The bolt cutter can be hard enough to use without having greased handles.

CUTTING/STRIKING TOOL HEADS

Flathead Axe

The procedure for maintaining the cutting edge of the axe is the same as above. The flathead axe striking surface requires that it be properly dressed. Inspect the striking surface for burrs, chips, or dings. Use the file to remove these surfaces, and round the surfaces slightly. The striking surfaces should be made slightly round to prevent these pieces of metal from forming during use. Don't overdo the rounding—just slightly. File the striking surface flat and clean. Deep pockets and creases will be formed while using the tool—these can't be prevented. Clean out these creases with sandpaper or steel wool to remove any loose metal or rust.

Once you've sharpened the blade and have dressed the edges, wipe down the tool head with light machine or motor oil to prevent rust.

Splitting Maul

Maintenance of the splitting maul is completely different from that of the axes. Removal of paint from the head of this tool is not required. Most splitting mauls have a rough finish, and the paint helps prevent rust. This tool's primary function is to split, so the paint will not really affect its splitting or cutting abilities. Remove paint from the cutting edge and striking surface. The cutting edge of the splitting maul is simply two opposing 45-degree angles. The splitting maul can be sharpened on a bench grinder or with a hand grinder. Grind the bevel to a 45-degree angle on both sides of the tool. A splitting maul's cutting edge is much blunter than that of an axe. Use a file and then a hone to remove the grinder marks, and sharpen the edge slightly. These marks are good places for rust to start.

Follow the procedures described below to care for the striking surface.

STRIKING TOOL HEADS

Maintenance of striking tool heads is primarily maintenance of the striking surface. Paint should not be removed from the heads of sledgehammers. Normally they have a rough finish to them, and the factory or user has applied paint to protect them from rust.

Check the striking surface for mushroomed metal. These pieces of metal are dangerous and should be removed. Use the file to remove any metal, and round the face of the striking surface slightly. Polish the striking surface to remove as many creases or dents as possible to prevent rust. Wipe the tool head with a light machine oil or motor oil to protect it from rust. Make sure that the striking surface and the cutting surface are oiled, especially since there is no paint there to protect the metal.

HANDLES

Wooden Handles

Wooden handles are high-maintenance items on tools. Handles used to be made of hickory or ash, but today they could be anything. If possible, use only hickory or ash for your tool handles. These woods are still available.

Any wood will rot, warp, chip, crack, and eventually break. Sunlight, pollution, water, and other natural and man-made substances attack wood. Inspect the tool handle thoroughly for any cracks or deterioration. If it is cracked or checked, replace it. If you have to glue or repair the handle to make it look functional, it will not *be* functional. Replacement handles are cheap. No amount of baling wire, wet leather wraps, staples, nails, or glue will salvage a damaged handle.

For routine maintenance, simply inspect the handle. Look for defects, dry rot, cracks, checks, or other problems that may be developing. Don't routinely sand and oil the handle. Repeated sanding will just plain wear out the wood by making it too thin. If a defect is found, or if the handle has been used a lot and looks as though it needs maintenance, sand it with a medium, then a fine-grade, sandpaper. The finish should be as smooth as a baseball bat. (Remember to remove any tape or French hitching before you start.) Use a tack rag to remove all the sanding dust.

Coat the handle with a liberal amount of boiled linseed oil. Use only boiled linseed oil. Work it in with your hands. Make sure that you put some on the top of the handle where it passes through the tool head. Wood is porous, and the oil needs to be worked into the wood. Coat the handle as evenly as possible. Wipe off any excess with a rag. Set the tool aside, and let the oil dry for a few hours before reinstalling any tape or French hitching.

Wooden handles shouldn't be painted at all. Under no circumstances should a working tool handle be varnished! Company markings may be added, but keep them to an absolute minimum. Paint hides defects in the handle and causes it to be slippery. Boiled linseed oil won't penetrate the paint, and the wood below the paint may dry out or rot.

Fiberglass Handles

Fiberglass handles with defects should be replaced. Chips, cracks, and abrasions are all serious indications that the handle has been stressed and probably needs to be replaced.

Overstrikes are the most common cause of damage to fiberglass handles. Overstrike protection will make them last longer.

Here's a quick fix for very minor abrasions to a fiberglass handle. First, thoroughly inspect the handle to make sure that a minor scrape is not an indicator of a serious problem. If it is truly minor, sand the spot with a fine grade of sandpaper until it is smooth. Apply a thin coat of polyurethane to the spot to protect the fibers in the handle. Fiberglass shards are razor sharp. Sanding them smooth and preventing them from coming up again by using polyurethane will make the handle last awhile longer. Fiberglass handles are inexpensive, and any damaged handle should be replaced. If the tool is going to fail, it won't fail during routine inspections. It will fail at a critical moment when you need it the most—when you need to squeeze that little extra out of it to get a task accomplished. Don't compromise your safety for a couple of bucks.

Plastic Handles

Tools, especially splitting mauls, do come with plastic handles. They're pretty strong, but they don't belong on your fire apparatus. Use a wood or fiberglass handle.

PRYING TOOLS

Prying tools need love, too! Although made of high-quality steel, prying tools are often used and abused. To maintain their peak effectiveness, maintenance is a must.

Pry Bar

Maintenance of this long piece of steel is relatively easy. Use a mill bastard

Although rarely used, pry bars must be maintained in a clean and oiled condition. This bar needs work.

file, single-cut or double-cut, to maintain the wedge point of the bar. Follow the original angle. The point should be chisel-like but not sharp. The tool should bevel back to allow the bar to get a good purchase but not to the point where the metal has been filed or ground so thin that it crumples or bends.

Use the file to dress the handle top of the tool also, especially if the bar has been struck with a striking tool. Dress the rounded end so that no metal mushrooming can be found. Make any sharp edges at the handle end slightly round.

To finish, sand it down, use a good primer, and paint. Don't allow too many coats of paint to accumulate before sanding down to bare metal and starting over.

Detroit Door Opener

Maintain this tool like a two-piece pry bar. Pull the sliding foot section out. Use a mill bastard file to sharpen the points on the foot so that they will penetrate the wood or dimple the metal of the door when used. Don't make them so sharp that they are a hazard—just sharp enough to get a good grab.

Make sure that the fulcrum point is well dressed. Follow the original contour and maintain that angle. If you have a well-maintained tool, use a small triangular file to sharpen the serrated teeth at the end of the fulcrum. Most likely, those teeth have either been ground off or have worn off most tools. If they are there, sharpen them up a little for a better purchase.

Lubricate the swivel of the foot pad with light oil, and lubricate all the swivel joints. Clean the sliding post that inserts into the foot pad bar. Clean the chain attachment for the clevis pin so that it moves freely. Also, clean the clevis pin so that it moves freely in any of the adjustment holes in the bar.

To finish, sand it down to bare metal, prime, then paint. Make sure the paint doesn't gum up any of the moving parts.

Claw Tool

Use a mill bastard file to maintain the point on the claw tool. This point is required to get the initial purchase for many of the functions of the tool. Taper the point to a reasonable diameter. If the point has been flattened out, use the file to reshape it. Don't make the point too thin or it will crumple under pressure and fail.

Dress the striking knob of the tool. Inspect it thoroughly for excessive damage, remove any mushroomed metal surfaces, and round the corners of the striking surface slightly.

Use the file to maintain the bevel on the bottom side of the fork. The fork should be somewhat sharpened but not razor sharp. Use a file and a hone. The fork bevels should be free of any dimples or flattened edges. Polish them as smooth as possible using a buffing wheel or other suitable means.

The tool can be painted, but it is easier to maintain if left without paint. A very light coating of oil on the shaft will prevent rust. Apply heavier (but not much) amounts of oil to the striking surface, point, and fork. If you are going to paint it, remove any old paint; then prime and repaint. Mask the bevels of the fork and the surface of the striking knob. These should be left as clean metal for better performance.

Kelly Tool

The kelly tool has several areas that require different types of maintenance. The head of the tool has the adze and the striking surfaces. Use a mill bastard file to maintain the bevel on the adze. This beveled edge should be sharp—again, not razor sharp, just sharp enough to shear metal bolts or screws. Use the file to remove any dimples or dings and to reshape the adze bevel. The bevel is underneath the adze section of the tool. Do not file a bevel into the top of the tool. Adding a bevel to the top edge of the tool will turn it into a wedge shape, and it will lose its advantage as a cutting adze.

Dress the striking surfaces. Remove all mushroomed metal, then file smooth and slightly rounded.

The chisel end of the tool should be sharpened somewhat like a splitting maul. Maintain the chisel edge by following the original opposing bevels. The edge should be sharp. File out any dimples or dings and reshape the edge. The angles on both sides should be equal.

Like the claw tool, the kelly tool can either be painted or left as raw metal. Oil it to prevent rust. Go lightly on the oil—don't make the tool so greasy that you can't hang on to it.

San Francisco Bar/Chicago Patrol Bar/Halligan Bar

These three tools are very similar in maintenance. As a matter of fact, with the exception of the pick on the halligan bar, their maintenance is exactly the same.

Adze. For the adze, use a good mill bastard file to maintain the bevel on the bottom side. Do not file a bevel onto the top of the tool. Maintain the original profile of the bevel. This bevel takes a lot of abuse, so don't file it too thin. The

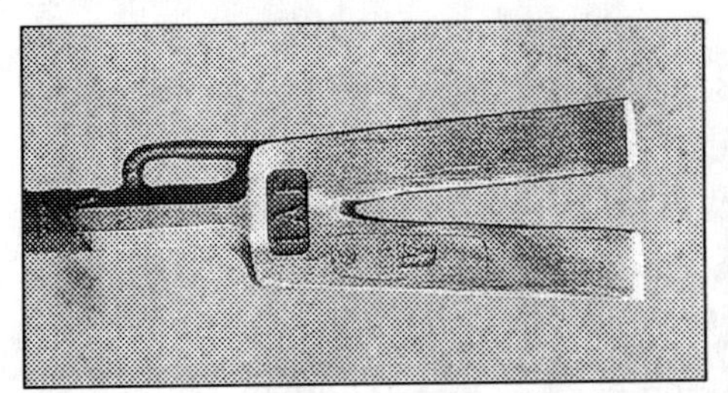
Keep the forks of these bars tuned up.

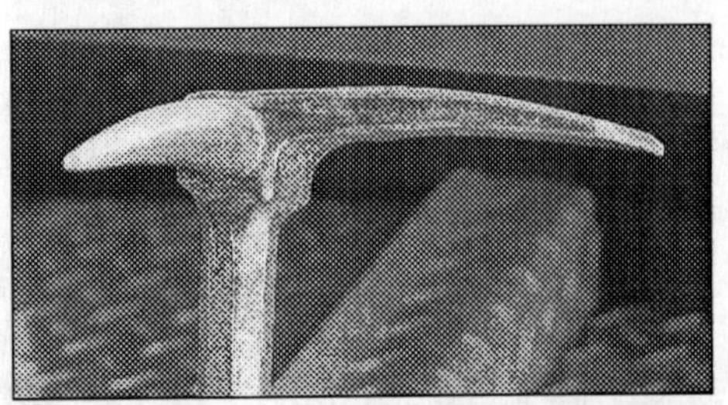
The adze and pick points must be kept sharp and beveled.

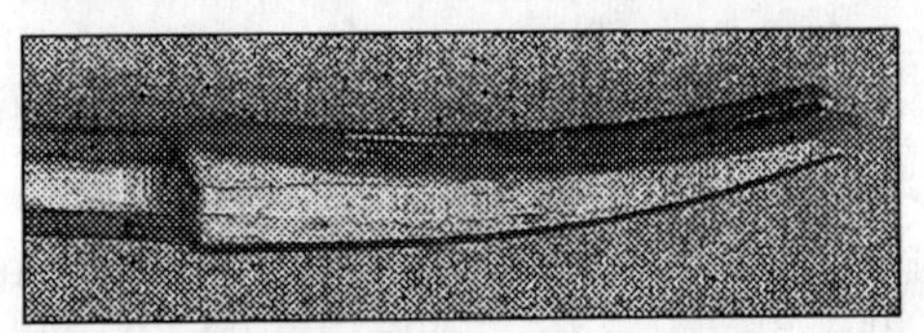
Maintain the gentle curve of the fork.

tool must be thick enough yet sharp enough for hard forcible entry techniques like shearing bolts and screws. Remove any curling, dimpling, or other defects in the adze edge, and reshape it with the file. Use a hone to sharpen the edge, being careful not to get the metal too thin. The tools are made of hard steel, so this is going to be work. Check and maintain the adze often. Case-hardened materials will knock the stuffing out of the adze end. When using the file on the adze, carefully maintain the slight curve of the adze top. Only remove dimpled or damaged metal at the end of the tool. Do not reshape the curve of the adze.

Striking surfaces. Dress the striking surfaces with the file. Remove all mushroomed metal. Do not overfile the tool; just remove the mushroomed stuff. Too many file marks in the tool will make it more prone to rust.

Hexagonal shaft. Maintain the hexagonal shaft by keeping the edges dressed. When you use it as a striking tool, a miss can sometimes cause the metal of the handle to mushroom. This won't happen often (I hope), but check it just the same.

Fork. Use the file to maintain the bevel on the bottom side of the fork. The fork should be somewhat sharpened but not razor sharp. Use a file and a hone. The fork bevels should be free of any dimples or flattened edges. Polish them as smooth as possible using a buffing wheel or other suitable means. Be careful to maintain the curve of the fork. Also, don't file the tips of the fork too thin. Maintain the original bevel.

Half round (San Francisco bar). Maintain the half round so that it is just that—half round. Flat spots should be filed back to half round again. Dress the edges of the half round to get rid of the sharp mushroomed metal.

Pick point (halligan). Keep the pick point of the halligan tapered and sharp—very sharp. Use the file to taper the point, then a good grade of wet/dry sandpaper for metal to smooth the entire pick point.

Snap hook attachment. If you have added a chain link to the tool for attaching a utility rope for venting windows from upper floors or the roof, inspect the link thoroughly for cracks in the weld. If the tool was engineered with this feature, it should still be checked. If the link has been damaged, file it off. Either replace it with another link or do without it. Throwing the tool off a roof with a damaged link is extremely dangerous!

PIKE POLES

Wooden Pole Handles

Wooden pole handles require a lot of work to be maintained. Usually poles are not carried inside the apparatus but are mounted on racks or in holders. This exposes them to weather, road dirt, salt, and other factors that will deteriorate them very quickly.

Inspect the pole for any cracks, chips, checking, abrasions, dings, or other problems. Severe dings or gouges will severely limit the use of the pole, so replace the handle. Minor dings or abrasions can be sanded out.

Use first a medium-grade sandpaper, then a fine sandpaper. Don't sand the handle so smooth that it is slippery. Allow a little of the grain to stand out to add (albeit very little) some grip. Carefully inspect the pole for any embedded objects like glass shards or small fragments of metal.

Once the pole has been sanded, use boiled linseed oil to coat the handle. Use your hands to apply it, and work it into the grain of the wood. Use only boiled linseed oil. Allow the pole handle to dry. Check the tool head for firm attachment to the pole.

Paint the tool head but not the pole itself. Never paint the handle of a pike pole. Company markings should be kept to a minimum size.

Fiberglass Poles

Fiberglass poles with defects should be replaced. Chips, cracks, and abrasions are all serious indications that the fiberglass pole has been stressed and probably needs to be replaced.

Overprying and striking window frames or metal framing are the most common causes of damage to fiberglass poles, especially up close to the head. Fix abrasions as described above (see page 149).

Metal Poles

Sand these poles down, prime them, and paint them. Bent or misshapen poles should be replaced. Minor bends can be straightened, but you will be stressing the metal even more. If a metal-handled pole is bent, it has been severely misused. Question its strength. Replace it.

PIKE POLE HEADS

National Pike Pole

Keep all the edges of the hook and pike well dressed and square. The square edges will help to penetrate most materials. The pike should be sharpened to a point. Use a hone to clean and sharpen the edges. Paint the tool head to help prevent rust.

Plaster Hook

Use a mill bastard file and a hone to sharpen the edges of the main pike point and the collapsible wings of the tool. These edges should be sharpened like an axe—that is, tapered to an edge that will cut through and not stick to the material. Oil the collapsible wings so that they move easily. Painting the head will help prevent rust, but be careful not to overpaint the head, or the wings won't move easily.

Chicago Pike Pole

Follow the original bevels of the tool head and maintain all the ridges with a file. Do not sharpen the edges; just maintain them. Sharpen the pike so that it will penetrate materials easily. Sharpen the edges of the hook, and file the downward point sharp, which will help the tool grab material with less effort. Don't file these areas too thin or too sharp, but maintain them sharp enough to get a purchase into wood or plaster.

To paint the tool head, sand it down with sandpaper or steel wool, prime it, then paint it.

New York Pike Pole and Arrow Hook

These pole heads are a pain to maintain. A big advantage of them is their massive size and weight. Other than a good cleaning after a fire, they don't usually require too much maintenance.

A mill bastard file should be used to maintain all of the edges of the tools. The edges, or ridges, help them penetrate material. Maintain the pike point by filing it sharp on two opposing sides. Don't file it too thin. It should be just a little sharper than blunt.

Clean the head well. Plaster and drywall really stick to the head. Use a wire wheel or stiff brush to get all the stuck material off. Prime and paint the head.

San Francisco Pike Pole

The San Francisco pike pole head has a lot of edges that need to be maintained. The pike head needs to be filed to a sharp point. Maintaining the pike to an arrowlike point allows the tool to penetrate wood, lath, plaster, and other materials more easily. Use a mill bastard file. Slowly, evenly file on both sides of the pike. File it down to suit your own eye and the type of use you will have for the tool.

Clean and well-maintained tools are the hallmark of a professional firefighting company.

Use the file again to maintain the angle on the top of the hook. This angle also helps penetrate material and allows the material to fall away from the hook head on the upstroke.

The San Francisco pike pole has a series of teeth along the bottom side of the hook. Carefully use a triangle-shaped file to maintain a very slight edge on the teeth and to clean out any impacted material stuck in there. The teeth need to be able to grab material. Easy does it on the filing—excessive filing will thin out the hook and make it more susceptible to breaking or bending. The hook is not a serrated knife, and it doesn't have to be that sharp.

Halligan Hook/Roofman's Hook/Multifunction Hook

The key to maintaining these tools is to maintain the bevels and the angles of the adze. Use a mill bastard file and a hone to keep their edges sharp. The ends should be as sharp as those of cutting tools.

Maintain the back angle of the bevel and the straight-edged bottom. File along the original angle to maintain it. Don't change the angle at all. Use the hone to remove all file marks and to put an edge on the adze.

Also, file along the tool wherever there is an angle, especially on the halligan tool head. The top point and triangular edges must be maintained to provide maximum penetrating power. Flat surfaces should be filed and honed flat.

To finish, sand the tool heads clean, then prime and paint.

Drywall Hook

Maintaining this tool is actually easier than it looks. Using the file, maintain the beveled cutting edge and point on the top fin of the tool. If the point has been flattened, reshape it. After filing, use the hone to put an edge on the fin. If you use this tool in metal frequently, this maintenance will be required often. Gypsum board won't damage this tool too much.

Use a triangular file, and file the teeth of the rake to a point. Make the teeth sharp enough to grab both gypsum board material and the paper. A few strokes is all it will take. The teeth need to be sharp but not razorlike. Using the triangular file will also help to clean any impacted material out of the teeth. If you have a small hone, hone the teeth. This will help take out file marks and thwart rust.

To finish, sand the tool head thoroughly, and use steel wool to get it good and smooth. Prime and paint the head.

EK Hook

Keep the EK hook sharp! This tool is designed for metal cutting. Use the file to maintain the cutting bevel along the top surface of the tool. Use slow and deliberate strokes to follow and maintain the tool's original cutting edge. Make sure that you remove any dimples or dents along the cutting edge. Hone the cutting edge sharp.

Maintain the heavy, square pulling edge on the bottom side of the tool. Keep the edges on the bottom (except the pulling teeth) square. If you have no cutting edge, the tool will be ineffective at cutting through material on the downstroke. The undersurface should be as sharp as the top cutting surface. File the teeth of the tool sharp using a triangular file. The teeth also need to be sharp to grab and cut through metal.

Clean the tool head thoroughly. Mask off all the cutting surfaces, then prime and paint the tool. Remove the mask from the cutting edge, then oil the edge with light machine oil or motor oil to prevent rust.

Boston Rake

Take a mill bastard file and maintain all of the edges of the rake. The edges should remain square. Remove any mushroomed metal. File the leading angle of the tool to maintain the point. Hone the leading edge of the tool to remove file marks. Sand, prime, and paint to finish.

Clemens Hook™

File the bevel on the bottom of the crescent head. Maintain the original angle of the bevel. Use the hone to remove file marks and sharpen the edge of the bevel. Use a small hone to maintain an edge on the cutting fin inside the crescent-shaped head. A small file may be needed to file out dents and dings. Use a hone on the fin to maintain a cutting edge.

File the pike of the head to maintain the small, sharp point. Don't get too carried away with the file or you'll misshape the head. Dress the edges of the tool to remove any mushroomed metal. Sand, prime, and paint to finish.

L.A. Trash Hook/Arson Rake

Maintain the points on the tines using a file or sandpaper. The points will dull quickly when the tool is used on a hard surface such as pavement or a garbage dumpster. The sharper the points, the better the grab. Don't go nuts making the points like razors, but keep them sharp enough to grab wood and soft metals. Sand, prime, and paint to finish.

Gatorback Hook and Dragonslayer™

Maintain these tools as you would a handsaw. File any dings or dents from the teeth. Because the teeth are easily bent, you may have to lay the tool on a hard, flat surface and tap them back into line. Maintain the edges of the teeth with a file. A triangular file works best for getting at them. A small hone should be used to maintain the cutting surface of the teeth. On the gatorback hook, maintain the pike pole head exactly the same as a national pike pole. Sand, prime, and paint to finish.

PERSONAL TOOLS

Officer's Halligan Hook

See Halligan Hook.

Officer's Tool (O Tool)

See A Tool.

Small Halligan Bar

See Halligan Bar and A Tool.

Truckman's Tool

See A Tool.

Fencer's Pliers

Maintain all of the cutting edges of the fencer's pliers with a small mill bastard file and hone. The cutting edges should be kept very sharp. Use wet/dry sandpaper to maintain the pick, and touch up any nicks or dings with a file. Use a light machine oil to prevent rust.

SEVERAL-IN-ONE TOOLS

T-N-T Tool

See Cutting/Striking Tool Heads and National Pike Pole. Use a mill bastard file to maintain the bevel on both sides of the chisel point of the tool. Hone the chisel very sharp, removing all of the filing marks. To finish, mask the striking surface and the chisel surface, then prime and paint. Remove the masking material and oil with a light machine oil or motor oil.

Cincinnati Tool

See Axes, Halligan Bar, and National Pike Pole.

Pry Axe

See Axes, Halligan Bar, Detroit Door Opener (for sliding bar).

Hux Bar

Maintain the edges of the nail puller using a file. Filing will remove the chrome, so the tool will rust unless it is coated with a light machine oil or motor oil. Use sandpaper to clean out and remove rust from any dings and creases. Damaged tools should be removed from service. Use a file to maintain the hydrant openings to their proper dimensions. This tool should be removed from service and replaced with a more useful implement.

SPECIAL-PURPOSE TOOLS

Bam-Bam Tool

The big maintenance item on the bam-bam tool is the screw chuck. Keep the threads well oiled and clean. If they get beat up, use an appropriate rethreading tool to repair them. Replace the screw after each use. Always have a large supply of screws available.

Keep the tool free of rust, and ensure that the slap hammer moves back and forth easily. Chromed tools are easily maintained unless you knock off some of the chrome. Stainless steel bam-bams are almost maintenance-free. Regular steel tools should be sanded, primed, and painted.

Hockey Puck Lock Breaker

Maintain this tool as you would any large pipe wrench. First, use a triangular file to keep the jaw's teeth very sharp (after all, the tool is expected to bite into case-hardened metals). Maintain the gear mechanism for the jaw adjustment by keeping it clean and well oiled. Paint most of the surfaces except the jaw areas, which should be lightly oiled to prevent rust. Thoroughly inspect the cheater bar attachment if you have a commercially made lock breaker. Make sure there are no weak points in the handle and that the resin glue is intact and holding well. If any defects are spotted, remove the tool from service and either have it repaired professionally or by someone who fully understands the complexity of resin glues and epoxies.

A Tool

The A tool must be maintained to function well. The cutting edges inside the A shape must be kept extremely sharp. Use a file and a small hone to maintain keen cutting edges on the tool. The blades must be able to bite hard into a lock cylinder, so sharpen them accordingly. There should be no dents or dings in the blades. Be careful to maintain the original bevel.

Just as important as the cutting blades are the very sharp points at the bottom of the A tool. These must also be maintained. The purpose of the A tool is for it to be driven in and behind flush-mounted or protected lock cylinders. The sharp points help guide the tool as you drive it in. Use a file to maintain those points. Keep them very sharp. They will wear down just by storing the tool and carrying it around.

J Tool

Maintain the tool in the shape of a J. Keep it rust-free and lightly oiled.

K Tool

The K tool's blades are removable. These should be kept as sharp as possible and free of all nicks and dings. A gouge in a blade may make it ineffective in pulling a lock cylinder. Use a file to shape the bevel, then a hone to sharpen it. These blades are driven into metal, so they need to be extremely sharp. If the blades are badly damaged, order replacements from the manufacturer. Keep the K tool clean and free of rust. A light coating of machine oil or motor oil will help. The tool can be painted, but don't paint the blade surfaces.

Duck-Billed Lock Breaker

Use a file to dress the edges of the tool, especially the striking surface. All mushroomed metal should be removed. Use a grinder or a file to maintain the profile so that the tool fits into the lock shackles of most locks. This may require tapering the duck-bill point, but don't file or grind it too thin. Maintain a good thick surface but one that is small enough to fit into the lock. Paint the tool.

Shove Knives

If they work, they're maintained! If they don't work, make new ones.

Vise Grip and Chain

Maintain this as you would a normal set of vise grips. Keep the screw assembly clean and oiled. Inspect the weld where the chain attaches to the vise grip. If cracks are noted, grind off the link and install a new one. Keep the vise grip and chain lightly oiled to prevent rust.

Battering Ram

Keep the edges dressed, remove all loose metal flakes, make sure the handles are tight, and dress the striking surface with a mill bastard file. Sand, prime, and paint.

Combination Punch and Chisel

Maintain this tool as you would any standard punch and chisel.

Hammerheaded Pick

Keep all of the surfaces well dressed. Maintain the handle (see Wooden Handles). Keep the striking surface clean and dressed, the edges slightly rounded. Square up and maintain the shape of the pick. The pick point does not need to be sharpened. Most often, it is better practice to blunt the pick slightly. Sand, prime, and paint the tool head. Do not paint the handle.

Rebar Window Breaker

Keep it painted and rust-free.

CHAPTER 10: TOOL COMBINATIONS

There is no such thing as one device that can accomplish all of the tasks a firefighter might encounter on the fireground. Such a tool does not exist. Firefighters perform many of the same tasks over and over at fires. Although no two fires are the same, the common tasks we perform are rescue, ventilation, forcible entry, and overhaul. All of these tasks require tools; many of them require different tools or combinations of them.

In this chapter, we will look at tool combinations. What two tools can be put together to perform most of the basic tasks required? Yes, firefighting is a science, but it isn't rocket science. Firefighters have been honing the art of firefighting for almost 400 years in this country alone. There isn't too much new; it just gets recycled. The tool combinations discussed in this book are the ones that have worked in hundreds of thousands of situations. They've been tested. Firefighters stake their lives on these tools every day.

A tool's efficiency depends on you. Size-up is critical to tool selection, and it should be made long before you respond to an alarm. Knowing the types of building construction in your response area is an absolute must. Know your town's history. By studying the history of your community, you will learn what types of buildings you may encounter and what building techniques were used. It is very important. Germans built buildings differently from the Swedish, who built differently from the French, who built totally differently from the colonists. Our own fire service history has played an important role in the way buildings are constructed. Huge conflagrations and high death tolls brought about the enactment of building and fire codes. As a firefighter, you are

Tools often need to be paired up. This Chicago firefighter has both a six-foot hook and an eight-pound pickhead axe.

responsible for every structure in your community—not just those under construction today, but every building ever built and still standing!

Become familiar with what types of security devices are being sold and used. A trip to the local home improvement store will answer a lot of questions about what types of tools you will need to bring with you to the next fire. Do prefire inspections. Don't just look for code violations—look at how the occupant secures the building at night. How do you get in? What tools do you need?

Standing in front of a locked door with the most sophisticated tools in the world won't open that door. Know the capabilities and limitations of the tools you are holding in your hand.

THE IRONS

The tool combination "the irons" means different things to firefighters depending on their geographic location. A set of irons is basically a prying tool married to (carried with) a striking tool. The term irons has long been a slang term for metal pry bars. The maritime industry, railroaders, and construction workers all have tools they call irons. In New York City, the original irons meant a pry bar plus a New York lock breaker. The irons man carried both tools, sometimes with a rope sling to carry them over his shoulder so that he could climb ladders or fire escapes. As tool innovations progressed, the types of tools carried as irons changed, but the name stuck.

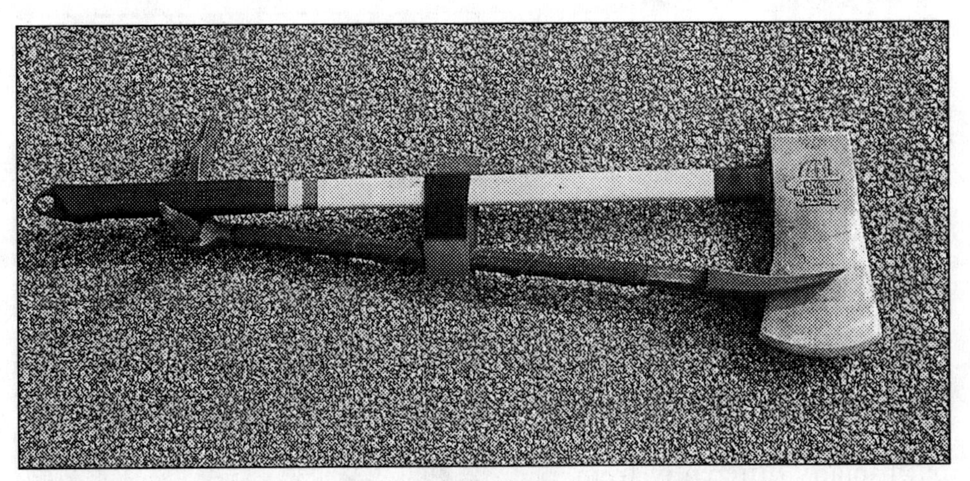

The irons—an eight-pound flathead axe married to a halligan bar.

Today, a set of irons is a flathead axe married with a halligan bar. The axe of choice is the eight-pound flathead. A standard 30-inch halligan bar will marry together almost perfectly with an axe. Stand a flathead axe on the floor, head down, and take a 30-inch halligan bar and slip the fork over the blade of the axe, the adze end pointing toward the axe handle. Lean the halligan bar forward so that the axe handle nests in the angle of the adze and the pick of the halligan bar. Grab both the halligan bar and the axe handle about midway down, or closer to the adze end of the tool if you have a small hand. That is a set of irons.

When carrying this set of irons, hang on tight—sometimes the halligan bar will slip. To improve carrying the set, use a strap to hold them together. There are some commercially available straps for that, but old rubber tourniquets work great. Nylon straps also work and can be cut down to fit the tool set perfectly. Carry the tool with the axe head next to your leg, blade down, and the halligan's hook out in front of you. The hook is the most dangerous part of the irons, and you want to know where it is at all times.

This firefighter is well prepared. He safely carries a six-foot hook and a set of irons.

Another type of irons can be made by marrying together an eight-pound splitting maul and a halligan bar. This combination is just as effective as the eight-pound flathead combination, except that the splitting maul is more capable of functioning as a sledgehammer because it has a sledgehammer face. The force concentration on the striking face of the splitting maul is better than that of the flathead. You do lose something using this combination, though: The splitting maul is not as efficient at cutting as the axe. The blade is fatter and the head is thicker. Putting an eight-pound splitting maul and a halligan bar together would be beneficial when responding to areas where smashing holes through roofs or walls is easier and faster than cutting. Many modern private dwelling roofs can have a hole smashed in them with a maul faster than they can be cut with an axe.

Irons can be a multitude of tool combinations. A favorite Midwestern tool combination is an eight-pound splitting maul and a small halligan bar. This tool combination is held together by an old section of 2½-inch hose. To make the hose holder, get an old section of 2½- or three-inch hose. Cut the outer jacket off the hose and slip it over the handle of the splitting maul. Slide the small halligan bar, fork first, down into the hose alongside the handle of the maul. You can now carry both tools easily.

One of the nice aspects of this tool combination is that the small halligan doesn't have to be removed to use the maul. By grabbing the hose and holding on, the maul can be swung to open lightly secured doors or to drive other tools. The tool must be removed if the maul is to be swung with any force, though.

Another advantage of the small halligan is that it can be used in confined areas. This tool combination cannot be used where high security is found, however, because the small halligan doesn't have the leverage of the full-size bar.

Older tools can be restored and returned to service as backup sets of irons. Tools that are old are only that—old. They will still perform if you know how to use them.

Some of the uses for a set of irons include:

- Conventional forcible entry through doors and windows.
- Forcible entry through padlocks and chains.
- Forcible entry through burglar bars and metal window gates.
- Forcible entry into automobiles.
- Forcible exit during search and rescue operations.
- Breaching walls.
- Cutting ventilation holes in floors and roofs.
- Removing skylights, scuttle covers, and roof caps.
- Opening elevator doors.
- Overhaul operations, including water removal.
- Footholds on pitched roofs.
- Door and window removal.
- Search tools.
- Utilities control.
- Anchor points for rope.
- Emergency bracing.
- Salvage tools.

LOCK-BREAKING COMBINATIONS

Tool sets can be put together to perform specific tasks. Marrying a striking tool and a lock-breaking tool is a fast and efficient way to deploy them when required. Locks can present significant obstacles to firefighters. Sets of irons can be expensive, and many fire departments only have one, maybe two sets at the most.

Lock-breaking sets can be put together very inexpensively. A duck-billed lock breaker and an eight-pound sledgehammer make a very effective pair. The duck bill can be manufactured in the firehouse from scrap metal, and the sledgehammer can be purchased at a hardware store for less than $20.00. This tool combination is held together by an old section of 2½-inch hose as mentioned above.

Older, less frequently used tools can also be revitalized and married together with striking tools to make backup sets of lock-breaking tools for whenever the irons have already gone to work. The claw tool and sledgehammer are one good example, as are the hammerheaded pick and sledgehammer or flathead axe.

TO THE ROOF

Firefighters who operate on the roof must be very experienced and capable.

Roof operations are among the more dangerous functions performed at the scene of a fire. Tools play a critical role in the success and safety of this operation.

Use tools whenever possible. Using just your hands or your body to perform certain tasks is risky.

Today, fire departments depend on power tools to speed up roof operations and to get the building ventilated as quickly as possible. Chain saws, circular saws, and other gasoline- or electric-driven tools do make fast work of cutting—when they start, that is. Hand tools always start, and firefighters operating on the roof with power tools must back them up with hand tools.

Exactly which tools should be taken to the roof will depend on the size-up you perform. What kind of roof is it? What's the pitch? What's the roof made of? Is it a truss roof? These questions must be answered before you can make your selection.

What task or tasks are you going to perform once you reach the roof? An example might be that the roofman will ventilate the roof, check it and adjoining exposures for fire extension, check all sides of the building for victims, move down from the roof, and perform a search from the top down.

A wide variety of hand tools may be needed during roof operations.

These basic functions from the roof require specific tools. Tools that you want at a minimum to perform those tasks include:

- A cutting tool (axe, splitting maul).
- A 12-foot push/pull tool (national pike pole, halligan hook, trash hook).
- A prying tool (halligan, Chicago patrol bar, claw tool).

You may need additional equipment such as roof ladders and other tools. With those tools you can ventilate the roof by cutting a hole with the axe, removing the debris with the pike pole. The pike pole would be inverted and inserted into the hole in all directions as far as you could reach to push down ceilings. The halligan would be available to drive the pike or hook into the roof for a foothold or to help in forcing open any skylights or ridge caps you may find.

Commercial buildings may call for a different set of tools. The function of the roofman will be the same, but the obstacles encountered will be different and more difficult. There may be bulkhead doors, shaft covers, scuttle covers, ventilators, HVAC systems, and other openings that may need to be forced open. The roof may still have to be cut.

The list of possibilities can go on and on. I don't have a single answer for what tools you must take to the roof every time. These are the basic tools, but only you can answer the question of what your exact needs are. Know your jurisdiction! Know the capabilities of your first-arriving companies and, most importantly, know how to use the tools that are available to you.

FORCING YOUR WAY IN OR OUT

During their training, firefighters spend a lot of time learning forcible entry techniques. Unfortunately, most don't get a chance to actually practice or use those skills they learn. Forcible entry props are few and far between. Videotapes help, but they don't give you the actual feel for dealing with all the variables that you may encounter during forcible entry.

Security in America has made entry into most homes and businesses a real obstacle for firefighters. Urban area firefighters are accustomed to the problem, but more often than not suburban and rural firefighters are surprised and often stumped.

Forcible entry can no longer be performed with a boot or a shoulder. Firefighters must be proficient with tools to gain access for rescue and firefighting. Proper tools are the solution to the forcible entry problem—there are no gimmicks. There are two basic types of forcible entry: (1) conventional forcible entry using a striking tool and a prying tool, and (2) through-the-lock forcible entry using special lock pullers and tools to trip and operate the lock normally after the lock cylinder has been removed.

Proficiency in these skills is critical to your safety. There are many situations in which you will have to force your way in to save a victim. There are equally

(Left) Always have a tool with you. You never know when you may have to force your way out of a fire building. (Right) Having hand tools with you will save your life. This Chicago firefighter is forcing a door that is blocking the exit for an engine crew. The engine crew had no tools with them.

as many, if not more, situations where you will have to force your way out to save your own skin. Understanding how tools work and what they are capable of is not just a way to kill time during drills—these are life skills.

Tools should not be left at the entrance door to a fire building. They should be with you at all times—and not just any tool for the sake of carrying one, but a tool that will function well, and a tool you know how to use. It is critical that specific tasks and assignments be given to firefighters. Tools (or, more importantly, tool combinations) must be available inside a structure. A firefighter whose only tool is a six-foot pike pole will perform well at the overhaul stage, but it will be useless to him if he must escape and encounters a locked or blocked path. By carrying a hook and a prying tool, he'd be better prepared to perform more than one task.

Personal tools are a great asset to you as a firefighter. A small pry bar is easy to carry, and may save your life if you must open a door, breach a wall, or otherwise force your way out of a life-threatening situation.

The following combinations have been proven to be outstanding for forcible entry. These tools will get you in and, more importantly, get you out again.

- An eight-pound flathead axe and a 30-inch halligan bar.
- A 10-pound sledgehammer and a 30-inch halligan bar.
- An eight-pound splitting maul and a 30-inch halligan bar.
- A six-foot roofman's hook and a 30-inch halligan bar.

OVERHAUL

Overhaul is the biggest job we do. As firefighters, we learn overhaul on the job, at the scene of an actual emergency. Performing overhaul at a fire is good experience, but training must take place before we go out and practice in somebody's living room. More firefighters are injured during overhaul than any other stage of a fire. You're tired and you want to get it over with, take up, and go home.

Tools are the only way for a firefighter to perform effective overhaul. During overhaul, firefighters often work harder than necessary to accomplish the task. By becoming proficient at tool selection and use, you will perform faster and better and be less likely to miss hidden fires.

Almost all of the tools used during the initial fire attack will be used during overhaul. Each type of tool is capable of performing tasks that are essential for opening up and removing debris after a fire.

Cutting tools—Cutting tools can be used to remove moldings, trim, and doors. Cutting tools will allow you to open walls and floors to check for hidden fires. At the overhaul stage you are very tired. Don't work with cutting tools over your head. Limit your use of cutting tools to materials that are at waist level or lower.

(Top right:) An axe is a valuable tool for overhaul. (Bottom left) Knowing the proper techniques for using poles will make overhaul operations easier. (Bottom right) Overhaul would be impossible without pike poles. Using them efficiently makes overhaul more effective.

Prying tools—Prying tools are more efficient for removing moldings, trim, and doors. Since they are designed to pry, they can be used more efficiently and with less effort than cutting tools. Well-dressed, beveled edges will pry and remove material much easier than cutting tools. In some circumstances, the prying tool will easily cut material that must be removed. Don't work with the tools over your head—prying tools should be used for materials that are located between your shoulders and the floor.

Poles—Poles should be used to remove material in areas higher than your shoulders. Selecting the most efficient pole will allow you to open walls and ceilings, remove molding and trim, and perform a variety of other tasks. Poles should be a minimum of six feet long. Better yet are eight-foot poles. Short hooks lack leverage, and their minimal reach is a disadvantage. If you're using a four-foot closet hook to overhaul a closet, you'll have to be in the closet to make it work and will be pulling all the debris down on top of yourself. Get a six-foot hook and stand in the doorway of the closet. You'll still pull it all down, but you won't be under it. Work the tool so that the debris falls out and away from you, not on you.

Several-in-one—They're great, but most of them are too short to be effective pike poles. If you're using this type of tool, limit its use to areas at the level of your shoulders and below. Most several-in-one tools have sharp tools at both ends. Be very careful when using them in areas where other firefighters are working or moving around.

S.W.A.T.

S.W.A.T. comes to the fire service from the Chicago Fire Department. It stands for Special Wrenches And Tools. The idea is for companies to have kits available of all those special tools that you don't often carry with you but often have a need for on the fireground.

S.W.A.T. kits are carried on many Chicago Fire Department squads and truck companies. Some battalion chiefs have them in their buggies. The kit is usually in a canvas tool bag or other carrying arrangement that can be quickly deployed on the fireground.

A typical S.W.A.T. kit consists of the following tools:

- Duck-billed lock breaker.
- Mallet or small sledgehammer.
- A tool.
- K tool.
- Key tools (for through-the-lock forcible entry).
- Bam-bam tool and extra screws.
- Shove knives.
- Dental picks (for through-the-lock forcible entry).
- Assorted screwdrivers.

- Crescent wrench.
- Pliers.
- Extra cut nails (for door chocks).
- Elevator keys.
- J tool.
- Vise grip and chain.

The S.W.A.T. kit should contain all the tools best suited for your response area, grouping many of the forcible entry tools you need into one portable pack. The S.W.A.T. kit provides the special tools for performing forcible entry through padlocks, the through-the-lock technique, and other instances where you just need that one extra or special tool. The S.W.A.T. kit is an excellent idea when responding to smells-and-bells calls to get you into a building to check an alarm without doing any damage to doors or windows.

The S.W.A.T. kit also provides a central staging area for all those tools on your apparatus. Instead of having to go from compartment to compartment, the tools are all together. You can get one tool from the kit or take the whole kit with you. S.W.A.T. kits are not expensive to put together. Many of the tools can be made in the firehouse or purchased at a local hardware store.

CHAPTER 11: TOOLS ON FIRE APPARATUS

Every piece of equipment owned or operated by a fire department is a tool. Firehouses, helmets, coats, hoses, nozzles, and lights are all tools. The biggest and most important tool we own is our rolling toolbox—the apparatus itself. It's with this tool that we get all of our other tools to the scene of an emergency.

How often do we really give consideration to the hand tools in the big toolbox? According to the standards that govern the building and equipping of apparatus, an engine company is only required as a minimum to have one six-pound flathead and one six-pound pickhead axe, one six-foot pike pole or plaster hook, and one eight-foot pike pole. That is all that is required to be on the apparatus before it is placed in service.

I mention this only because some department and city officials take the standard as gospel, and those are the only hand tools the engine ever has assigned to it! Go ahead and laugh, but that is the truth in hundreds of fire departments across the country.

As you are reading this chapter, I hope that you can recall or refer back to all the previous chapters. Hand tools are critical to firefighting. There are no gimmicks and no magic. We are still using the same tools we have been using for years—sometimes a thousand years or more.

The tools you carry on your apparatus should be the result of the size-up you do of your community and your department. Determine what each piece of apparatus really does. Do your engines only function as engines, or do they go to car accidents? Do you have a truck company? If not, which apparatus carries the truck tools? Why is the tanker leaving the station with nothing on it but hydrant wrenches?

Ask tough questions and make a tough size-up. If in doubt, ask for help. Contact other local departments to see what they carry. Get tool catalogs and see what tools are available. Buy this book and find out how to use them!

The rest of this chapter will look at what hand tools should be carried on basic apparatus. This isn't written in stone. This isn't a standard. It's something to make you think.

ENGINE COMPANIES

For the purposes of this book, an engine company is a piece of fire apparatus whose main function is to respond to fires and other emergencies. At fires, the

Hand tools are as critical to an engine company as hoses and nozzles are.

engine is responsible for fire extinguishment. Its purpose at emergencies can vary from silencing alarms to performing vehicle extrication. It has a minimum crew of three firefighters.

Consider the following as a complement of hand tools for the engine company.

Cutting tools—At least two eight-pound pickhead axes, mounted in brackets for ease of accessibility by the firefighters. The axes will be required for ventilation and overhaul both during and after the fire. Not all fire is visible to the attack team, and the pickhead axes may be needed to open walls or floors to get at the fire. Two different sizes of bolt cutters should also be carried.

Cutting/striking tools—Two eight-pound flathead axes, to be married with two pry bars. One eight-pound splitting maul. These tools may be required if the engine company encounters forcible entry problems on arrival. They may also be needed at nonfire emergencies such as rescues and car accidents.

Prying tools—Two halligan bars should be married with the two eight-pound flathead axes and stored as two sets of irons. Additionally, two pinch or wedge-point bars should be carried.

Striking tools—One 10-pound and one 16-pound sledgehammer. Engine company officers may be required to breach walls made of both wood and masonry to get the best position to attack the fire with a hoseline. Lightweight tools take too long; the fire isn't going to wait for the one vehicle in your department that has a heavy sledge.

Poles—Two six-foot poles, two eight-foot poles, one 12-foot pole, and one 16-foot pole should be carried. Another choice might be two six-foot poles plus one eight-foot, one 10-foot, one 12-foot, and one 16-foot pole. Select the tool heads that will best suit the type of area to which the company responds. Overhaul is a critical part of fire extinguishment. Not all buildings today have standard eight-foot ceilings. Many homes are being built with 12- to 14-foot vaulted ceilings and other areas that are impossible to reach with a standard six- or eight-foot pole. With the recommended complement of poles, salvage operations will be ineffective without calling for more companies or additional tools.

Other tools—Building a S.W.A.T. kit for the engine will allow through-the-lock functions at fires. More importantly, the engine crew will be able to access buildings quickly for stretching additional lines into exposures, checking fire alarms, and assisting civilians for EMS-type runs. If a S.W.A.T. kit is not practical, at least one A tool should be carried. Mount a piece of PVC pipe to the officer's door and slide in an officer's tool.

This may seem ridiculous to some firefighters. A majority of the fire departments in this country run without the benefit of truck or squad companies arriving first, or even afterward, to assist. Many departments function with only two or three engines, plus a tanker and a utility vehicle.

The tool complement described above enables firefighters to function alone until help arrives, either from within the department or from mutual-aid companies. The tool complement also allows the engine to function at a multitude of different emergencies.

Large-scale disasters deplete firefighting forces quickly. Floods, tornadoes, heat waves, thunderstorms, and blizzards have all required individual companies to function alone for indefinite periods of time. In some locations, companies may be cut off from the second-due unit because of flooded rivers or creeks during heavy rainstorms.

You must size up what that engine company is expected to do. Many engines respond to automobile accidents and other medical emergencies. All paramedics and EMTs are trained in CPR. When the defibrillator konks out in the middle of a code, the patient isn't lost. Without hand tools, what do you do when the power unit for the hydraulic tool quits? If you have a decent complement of tools, then the building isn't lost!

Tools should be stored or mounted so that they can be found and retrieved easily. Compartments used as tool bins should be organized and neat. A bin with the tools just thrown in is a waste of space and damaging to both the apparatus and the tools.

A plywood bottom can be added to the bottom of the tool compartment. Cutouts or brackets to hold each tool will organize the bin and make for easier selection during an emergency. Inventory and maintenance will be simplified. With organized bins, your tools are also less likely to be left at the scene. One look and it is obvious whether something is missing.

Poles can be mounted in upright brackets in walkways or along running boards. Otherwise, pole holders can be made out of 1½-inch or two-inch PVC pipe. By making a false bottom in the hosebed, these tubes can be installed and the poles stored underneath the hose load. The total loss of bed space would be no more than three to four inches.

Brackets can be installed on the insides of compartment doors or cab doors. This is an excellent place to mount axes. Otherwise, axes can be installed on the outside of the cab in the traditional location or mounted on the tailboard.

Locating and storing the tools is a problem, but it isn't insurmountable. After all, most of the tools we use were reinvented by firefighters. It won't take a group of firefighters long to figure out where to store them on the engine.

TRUCK COMPANIES

The engine company as described above is equipped to perform many of the functions of truck work at the scene of a fire but not all the functions of a truck company. Not all fire departments have truck companies. For this reason, engine companies must have a decent complement of tools to perform all of the tasks normally performed by a truck company.

Truck companies should be equipped with a large and varied selection of hand tools.

For purposes of clarity, a truck company is a piece of fire apparatus equipped with an aerial device. This device may include a ladder, aerial platform, aerial tower, or other similar device.

For fire departments that have a truck company, the complement of tools should be even greater than that of the engine. Even more importantly, their tools, tasks, and personnel should be appropriately assigned. All too often, fire departments have truck companies staffed by engine guys. The apparatus functions as another engine without any real thought being given to the requirements and tasks of a truck company.

Tools carried by truck companies should reflect the multiple functions and duties that the firefighters assigned to that apparatus must perform. Consider the following suggestions for tools to be carried by a truck company.

Cutting tools—At least four eight-pound pickhead axes, mounted in brackets for ease of accessibility. An assortment of different types of bolt cutters should be on the rig also.

Cutting/striking tools—Three eight-pound flathead axes. Two of the axes should be married with pry bars. These tools should be the basic tools for the forcible entry team of the truck. Two eight-pound splitting mauls should also be carried. As explained in Chapter 2, splitting mauls are excellent tools to take to a wood roof.

Prying tools—Three halligan bars should be available on the rig. Marry two halligan bars to two eight-pound flathead axes, and store them together as sets of irons. The third bar is available as a stand-alone pry bar, or it can be mated up with a 10-pound sledgehammer. Additionally, four pinch or wedge-point bars should be carried. Rescue situations may require the truck company to provide extreme leverage capabilities.

Striking tools—Two 10-pound and one 16-pound sledgehammer should round out the striking capabilities of the truck. Consider having one of the 10-pounders with a cut-off handle. Truck companies may be required to breach walls of both wood and masonry to gain access or to effect a rescue. These tools will be imperative if firefighters are trapped in a building collapse. Lightweight tools take too long.

Poles—Four six-foot poles, three eight-foot poles, two 10-foot poles, and one 12- and one 16-foot pole should be carried. Select the heads that will best suit the type of area to which the truck responds. Don't have a supply of tools all with the same head. If a particular hook is inefficient in a six-foot length, it will be even worse at eight or 12 feet! Overhaul is a critical part of fire extinguishment and a major assignment for the truck. Overhaul can be seriously complicated by tall ceilings, multiple levels of ceilings, illegal building modifications, and a host of other circumstances. Truck companies must be equipped to overcome any problems they encounter during forcible entry or overhaul situations.

Other tools—A S.W.A.T. kit is essential for a truck company. The standard for aerials recommends a specific toolbox to be carried by a truck. The S.W.A.T. is an addition to that toolbox. In addition to the S.W.A.T. kit, consider adding a hockey puck lock breaker.

SQUAD COMPANIES

Squad companies and rescue companies will carry more tools than a truck company. Most often, these will be specialized devices such as hydraulic tools, torches, and other technical and highly task-specific tools. Much of the apparatus will also be used as a command vehicle or haz-mat vehicle.

Tools for squad companies are really local issues. These companies perform specific tasks or augment regular line companies. If your department is considering starting a squad or rescue company, here is a basic tool complement suggestion.

Cutting tools—At least two eight-pound pickhead axes, mounted in brackets for ease of accessibility by the firefighters. Two axes are sufficient, since the company will need the extra storage space for its special tools. A pair of bolt cutters should be carried. Additional tools can be taken from the engine or truck.

Cutting/striking tools—Two eight-pound flathead axes, which should be married with pry bars. These tools will be for the forcible entry team of the squad. In many areas, squad or rescue companies function as supplemental truck companies or as rapid intervention companies for firefighter rescue. Two eight-pound splitting mauls should also be carried.

Prying tools—Three halligan bars should be available from the rig. Marry two halligan bars with two eight-pound flathead axes, and store them together as sets of irons. The third bar is available as a stand-alone pry bar or it can be mated up with a 10-pound sledgehammer. Additionally, two pinch or wedge-point bars should be carried.

Striking tools—Three 10-pound and two 16-pound sledgehammers will enhance the striking capabilities of the squad's crew. Consider having two of the 10-pounders with cut-off handles. Squad companies may initiate or assist rescues that require breaching masonry walls. These tools will be critical if firefighters are trapped in a building collapse.

Poles—Two six-foot poles, one eight-foot pole, one 10-foot pole, and one 12-foot pole should be carried. Select the tool heads that will best suit your response area. Don't carry tools that all have the same head.

Other tools—A S.W.A.T. kit is essential for a squad or rescue company. If the unit is to be mobile and will respond to all types of alarms, consider having two S.W.A.T. kits. If the squad should be called away before operations are finished at one scene, a kit can be left. A hockey puck lock breaker is an important tool for the squad.

THE CHIEF'S BUGGY

In addition to the Sanborn maps, clipboards, reams of documents, and reflective vests, the chief's buggy should also carry a small complement of tools. There are instances when the buggy arrives first, and without basic tools, nothing can be done. Many outstanding rescues have been made by the chief and his driver before the first-arriving companies even got there. Those chiefs had tools.

The complement of tools should fit the type of vehicle the chief is using. Consider these tools for the buggy:

Cutting tools—One eight-pound pickhead axe.

Cutting/striking tools—One eight-pound flathead axe.

Prying tools—One halligan bar married to an eight-pound flathead axe.

Striking tools—One 10-pound sledgehammer with a cut-off handle.

Poles—None.

The chief's buggy should carry a basic assortment of hand tools.

Other tools—A modified S.W.A.T. kit should be carried by the chief. At minimum, it should contain an A tool, a duck-billed lock breaker, key tools, elevator keys, and a pouch containing a variety of screwdrivers and pliers.